# Supply Chain Collaboration

Mei Cao · Qingyu Zhang

# Supply Chain Collaboration

## Roles of Interorganizational Systems, Trust, and Collaborative Culture

 Springer

Mei Cao
University of Wisconsin—Superior
Superior, WI
USA

Qingyu Zhang
Arkansas State University
Jonesboro, AR
USA

ISBN 978-1-4471-6215-5      ISBN 978-1-4471-4591-2 (eBook)
DOI 10.1007/978-1-4471-4591-2
Springer London Heidelberg New York Dordrecht

Printed on acid-free paper

Springer is part of Springer Science+Business Media (www.springer.com)

# Preface

## Objectives of the Book

The competition today is no longer between individual firms but between supply chains. To survive and thrive in the competition, firms have strived to achieve greater supply chain collaboration to leverage the resources and knowledge of suppliers and customers. Internet-based technologies, particularly interorganizational systems (IOS), further extend the firms' opportunities to strengthen their supply chain partnerships and share real-time information to optimize their operations. The objective of this book is to uncover the nature and characteristics, antecedents, and consequences of supply chain collaboration from multiple theoretical perspectives.

The book conceptualizes supply chain collaboration as seven interconnecting elements: information sharing, goal congruence, decision synchronization, incentive alignment, resource sharing, collaborative communication, and joint knowledge creation. These seven components in concert are necessary and sufficient to define the occurrence of collaborative efforts. This definition and its components allow us to explain supply chain collaboration more precisely. The book also operationalizes collaborative advantage as five components to capture the joint competitive advantages and benefits among supply chain partners.

The book applies multiple theories (e.g., transaction cost, resource based, social exchange, trust-based rationalism, knowledge-based theories, relational view, and extended resource-based view) to explain the role of IOS in supply chain collaboration. Grounded in extensive literature, the book proposes a theoretical framework relating supply chain collaboration, its antecedents (IT capability, IOS appropriation, collaborative culture, and trust) and its consequences (collaborative advantage and firm performance). Reliable and valid instruments of these constructs were developed through rigorous empirical and statistical analysis. The methodology employed includes structured interviews, Q-Sort, and a large-scale study. Data were collected through a Web survey of U.S. manufacturing firms in various industries and 211 usable responses were generated. The statistical methods used include confirmatory factor analysis and structural equation modeling (i.e., LISREL).

The book extends our understanding of the attributes of supply chain collaboration and collaborative advantage, the forces leading to the development of supply chain partnership, and issues involved in creating and managing the partnership. A better understanding of supply chain collaboration leads to the better management of it. The definitions and measures developed in the book allow us to better examine the important issues of (1) coordination, cooperation, integration, and collaboration, (2) joint value creation versus value appropriation, (3) common benefits versus private benefits, and (4) collaborative advantage versus competitive advantage.

## Who Should Read the Book

The book will be useful for anyone who is interested in supply chain collaboration, its antecedents, and its consequences. It includes people from multiple disciplines such as marketing, management, information systems, strategies, and operations management. It also includes students (mostly graduate students), researchers, faculty, and practitioners of supply chain collaboration and management. The book will benefit people from multiple business functions such as purchasing/procurement, manufacturing/operations, distribution/warehouse, transportation/logistics, supply chain management, and/or information technology.

One of features of the book is to use a variety of the tools for theory building and empirical research methods. In this book, we provide a more accurate and comprehensive conceptualization of supply chain collaboration. It is the first of its kind to operationalize collaborative advantage. It develops valid and reliable instruments of supply chain collaboration, collaborative advantages, and their related constructs through rigorous statistical methodologies including structured interview, Q-sort method, large-scale Web survey, confirmatory factor analysis, reliability, convergent and discriminant validity, validation of second-order construct, structural equation modeling, and multi-group analysis using LISREL.

## Organization of the Book

The book is based on Dr. Mei Cao's unpublished dissertation work in 2007. We revise and reorganize it by incorporating our published articles on this subject and the recent development in the field (with citation and reference list). The book is organized into eight chapters.

- Chapter 1 ("Introduction") gives an overview of supply chain collaboration and define key concepts and research questions.
- Chapter 2 ("Theory and Theoretical Framework") discusses ten different theories that can be used to explain supply chain collaboration and introduces the research framework.
- Chapter 3 ("Antecedents of Supply Chain Collaboration") defines and elaborates key antecedents of supply chain collaboration including IT resources, IOS appropriation, collaborative culture, and trust.
- Chapter 4 ("Supply Chain Collaboration Characterization") identifies and defines the nature and characteristics of supply chain collaboration as seven interweaving components of information sharing, goal congruence, decision synchronization, incentive alignment, resource sharing, collaborative communication, and joint knowledge creation.
- Chapter 5 ("Collaborative Advantage as Consequences") operationalizes collaborative advantage as consequences of supply chain collaboration with five elements of process efficiency, offering flexibility, business synergy, quality, and innovation.
- Chapter 6 ("Structured Interview and Q-Sort") describes the procedures for structured interview and Q-sort and reports the results.
- Chapter 7 ("Large-Scale Analysis and Testing") analyzes instruments and tests hypotheses through rigorous statistical methodologies including confirmatory factor analysis, reliability, convergent and discriminant validity, validation of second-order construct, structural equation modeling, and multi-group analysis in LISREL.
- Chapter 8 ("Research and Managerial Insights") offers additional discussion of the moderation effect of firm size and provides management insights and guidelines.

# Acknowledgments

We wish to thank our editors at Springer, Anthony Doyle and Fermine Shaly, as well as senior editorial assistants, Grace Quinn and Claire Protherough, and the rest of the editorial team at Springer, for bringing this project to publication.

We would like to thank our professors at University of Toledo. Special thanks go to Mark A. Vonderembse and T. S. Ragu-Nathan, who served as our advisors, for their intellectual guidance and support. Thanks also go to Ron Zallocco and Peter Lindquist for their thorough review of this research and helpful suggestions. Without them, this project would not be possible.

Our work has benefited from people who participated as subjects in this study and researchers in this area through their various publications. We owe a special debt to them.

Finally, we thank our extended families for their love, support, and encouragement during this project.

# Contents

# Chapter 1
# Introduction

**Abstract** Facing uncertain environments, firms (e.g., Hewlett-Packard, IBM, Dell, Procter and Gamble) have strived to achieve greater supply chain collaboration to leverage the resources and knowledge of their key suppliers and customers. Despite the benefits of supply chain collaborations, many of them fail to meet the expectations. It seems that collaboration has the most unsatisfactory track record of all supply chain management strategies. Supply chain collaboration seems to have great potential but further investigation is needed to recognize its value. In this chapter, we discuss research background, gaps in literature, research questions, and purported contributions. We also clarify some key concepts including cooperation, coordination, integration, and collaboration; joint value creation versus value appropriation; common benefits versus private benefits; collaborative advantage versus competitive advantage.

## 1.1 Background

With rapid changes in environments, advances in technology, and globalization of markets, organizations have become increasingly aware of the needs to optimize the performance of whole supply chains rather than individual organizations (Lambert and Cooper 2000; Lejeune and Yakova 2005; Nyaga et al. 2010; Fawcett et al. 2012; Verdecho et al. 2012). To survive and thrive in this emerging competitive environment, firms strive to achieve greater supply chain collaboration (Lee and Whang 2001) to leverage the resources and knowledge of their suppliers and customers (Fawcett and Magnan 2004; Verwaal and Hesselmans 2004; Lejeune and Yakova 2005; Malhotra et al. 2005; Cao and Zhang 2011), which may be the ultimate core capability (Sanders and Premus 2005). Prahalad and Ramaswamy (2001, p. 2) claim, "Being opposed to collaboration these days is a bit like being against quality, or maybe even profitability."

M. Cao and Q. Zhang, *Supply Chain Collaboration*,
DOI: 10.1007/978-1-4471-4591-2_1, © Springer-Verlag London 2013

Firms such as Hewlett-Packard, IBM, Dell, Procter and Gamble have forged long-term, collaborative relationships with their suppliers to achieve a stronger competitive position (Spekman 1988; Stuart and McCutcheon 1996; Dyer and Singh 1998; Dell and Fredman 1999; Parks 1999; Barratt and Oliveira 2001; Callioni and Billington 2001; Handfield and Bechtel 2002; Johnson and Sohi 2003; Liker and Choi 2005; Sheu et al. 2006; Cao and Zhang 2011). Scholars regard forming supply chain partnerships as an alternative to the traditional make or buy choice (Blois 1996; Kay 1997; Casson 1998) where partners develop idiosyncratic interfirm relationships through specific asset investment, shared know-how, complementary assets, and effective governance mechanisms (Williamson 1985; Gulati 1995b; Dyer and Singh 1998; Kaufman et al. 2000).

Supply chain collaboration means two or more autonomous firms working together to plan and execute supply chain activities (Simatupang and Sridharan 2002; Cao and Zhang 2011). Collaboration requires a certain degree of relationship among supply chain members (Lambert et al. 1998; Lejeune and Yakova 2005). It also requires supply chain members to share resources to meet their customer needs (Narus and Anderson 1996). Supply chain collaboration involves many coordination issues from different disciplines, such as customer relationship management (marketing), inventory, production, and distribution management (operations management), strategic alliances (organizational management), and electronic data interchange (EDI) and radio frequency identification (RFID) (information technology) (Croom et al. 2000; Lejeune and Yakova 2005).

Supply chain collaboration can deliver substantial benefits and advantages to its partners (Mentzer et al. 2000; Verdecho et al. 2012). Collaborative relationships can help firms obtain information (Gulati 1995; Koka and Prescott 2002), share risks (Kogut 1988), access complementary resources (Eisenhardt and Schoonhoven 1996; Park et al. 2004), reduce product development costs (Henderson and Cockburn 1994), reduce logistical costs (Stank et al. 2001), reduce transaction costs and enhance productivity (Kalwani and Narayandas 1995), improve quality (Newman 1988; Stuart and McCutcheon 1996), improve technological capabilities (Powell et al. 1996), enhance profit performance and competitive advantage over time (Mohr and Spekman 1994; Dyer and Singh 1998; Jap 1999; Mentzer et al. 2000). Without effective relationships, managing the flow of materials and information across supply chain are unlikely to be successful (Handfield and Nichols 2002; Lambert et al. 2004).

Internet based information and communication technologies (ICT), particularly interorganizational systems (IOS), further extend firms' opportunities to strengthen their supply chain partnerships and share real-time information to optimize their operations (Lejeune and Yakova 2005; Teo et al. 2011; Rai et al. 2012; Grover and Kohli 2012). Using IOS, supply chain partners can develop close relationships in the chain structure, which enables them to access each other's privileged data and information (Holland 1995; Rai et al. 2012). Such electronic hierarchies allow firms to achieve the effect of vertical integration without ownership through the use of IOS to tie-in partners and lock out competitors, and thus achieve sustainable competitive advantage (Konsynski and McFarlan 1990).

Firms have used IOS, such as EDI, to develop collaborative and long-lasting relationships with their supply chain partners (Son et al. 2005). IOS supports tightly coupled partnership that leverages capabilities of ICT, such as electronic integration (Venkatraman and Zaheer 1990), electronic partnership (Hart and Saunders 1998), and information partnership (Konsynski and McFarlan, 1990). IOS such as Collaborative Planning, Forecasting and Replenishment (CPFR), Vendor Managed Inventory (VMI), Efficient Consumer Response (ECR), and Continuous Replenishment (CR) takes supply chain collaboration from passive exchange of information between partners to proactive joint planning and synchronization of activities and business processes (Jagdev and Thoben 2001; Parks 2001; Skjoett-Larsen et al. 2003; Holweg et al. 2005).

While individual success stories of IOS (e.g., CPFR and CR) use in partnerships have been reported, mainstream implementation has been much less successful than expected (Holweg et al. 2005). Despite the benefits of supply chain partnering, many partner relationships fail to meet the participants' expectations (Niederkofler 1991; Hatfield and Pearce 1994; Doz and Hamel 1998; Barringer and Harrison 2000). It is widely observed that few firms are actually engaged in the level of integration that supply chain collaboration suggests (Fawcett and Magnan 2004) and few firms have truly capitalized on the potential of supply chain collaboration (Barratt 2003; Crum and Palmatier 2004; Min et al. 2005). As Sabath and Fontanella (2002, p. 24) note, "Collaboration arguably has the most disappointing track record of the various supply chain management strategies introduced to date." Supply chain collaboration seems to have great potential, but further investigation is needed to recognize its value (Goffin et al. 2006).

## 1.2  Key Concepts

Before discussing the research gap, there are several terms that need to be clarified including cooperation, coordination, integration, and collaboration; value creation and value appropriation; common benefits and private benefits; and collaborative advantage and competitive advantage.

### 1.2.1  Coordination, Cooperation, Integration, and Collaboration

Cooperation is to achieve unity of motivation (e.g., incentive alignment). In cooperative relationships, the firms might share information, expertise, cost, risk, and reward. Each firm remains completely autonomous from the others. They might change the way they work and learn from others.

Coordination is to achieve unity of action (e.g., decision synchronization) due to the interdependence. Although involved in coordination, firms retain control over their own operations. If each firm acts rationally and selfishly, the resulting outcome is generally suboptimal for the chain as a whole—the total benefits received by the chain are smaller than those if the firms could make their actions in line with one another (i.e., coordination) (Snyder and Shen 2011). Even the supply chain game moves to a Nash equilibrium—an outcome such that no firm can change strategies unilaterally and improve his own payoff, the outcome may not be Pareto optimal since some partners might be better off and some of them worse off (Snyder and Shen 2011). Thus an incentive alignment mechanism (e.g., buyback contract, revenue sharing contract, and quantity flexibility contract) should be in place to make the coordination work.

The coordination issue might persist even when cooperation (i.e., unison of motivation) has been attained. Actions cannot be aligned when firms do not recognize that their actions are interdependent with others. The problem can be solved through synchronizing decisions, establishing routines, procedures or rules for different contingencies, and enhancing knowledge of interdependence with others (Thompson 1967).

Besides cooperation and coordination, an organization has to achieve goal congruence with their supply chain partners, and then it can be said well integrated or collaborated. Integration and collaboration sometimes have been used interchangeably since both refer to a tight coupling process between supply chain partners. However, the term integration means the unified control (or ownership) of several successive or similar process formerly carried on independently (Webster's Third New International Dictionary 1966; Flynn et al. 2010). So it puts more emphasis on central control, ownership, or process coupling (e.g., EDI) governed by contract means. According to transaction cost economics theory, between the two ends of governance continuum of vertical integration and market exchange, collaboration is an intermediate form of hybrid governance. Collaboration is attractive since it puts more emphasis on governance through relational means in addition to governance through contract means (Nyaga et al. 2010). Thus supply chain collaboration is a better concept to capture the joint relationship between autonomous supply chain partners. The collaboration is not to have one partner to get a smaller piece of the pie so that others can obtain a larger piece, but it is to make the pie larger so that all partners can get larger piece than they had before.

## 1.2.2 Joint Value Creation Versus Value Appropriation

Value creation involves the total net value created in a collaborative relationship among partners. Value appropriation describes the net value that a firm gets from the collaboration. Thus value creation and value appropriation stand for the two sides of the same coin and they concern about the size of the total value pie created and the value slice appropriated (Lavie 2006; Adegbesan and Higgins 2011; Coff 2010; Wagner et al. 2010).

### 1.2.3  Common Benefit Versus Private Benefits

According to Khanna et al. (1998), common benefits are those that accrue to each partner in collaboration from the collective assimilation and application of the learning. These benefits are gained from operations that are related to the collaboration. Private benefits are those that a firm can produce unilaterally by assimilating skills and learning from its partners and applying and internalizing them to its own operations that are typically unrelated to the collaborative activities.

Common benefits mainly come from the creation of joint value through sharing of resources while private benefits derive primarily from appropriation hazards. Pursuits of common benefits and synergies encourage collaborative behavior while private benefits promote non-collaborative behavior (Kumar 2011). According to the relational view and extended resource based view (Dyer and Singh 1998; Lavie 2006), supply chain partners can work closely together to create common benefits (i.e., relational rents) and thus create collaborative advantage, a strategic benefits more than those by any firm acting alone.

### 1.2.4  Collaborative Advantage Versus Competitive Advantage

Supply chain collaboration is rooted in a paradigm of collaborative advantage (Kanter 1994; Dyer 2000) rather than competitive advantage (Porter 1985). According to the collaborative paradigm, a supply chain is composed of a sequence or network of interdependent relationships fostered through strategic alliances and collaboration (Chen and Paulraj 2004). Collaborative advantage comes from relational rents that produce common benefits for bilateral rent-seeking behaviors while competitive advantage encourages individual rent-seeking behaviors that maximize a firm's own benefits (Lavie 2006). The perspective of collaborative advantage enables supply chain partners to view supply chain collaboration as a positive-sum game rather than a zero-sum game where partners strive to appropriate more relational rents for their own competitive advantage (Cao and Zhang 2011).

Collaborative advantage is a relational view of interorganizational competitive advantage (Dyer and Singh 1998). Collaborative advantage comes from relational rent, a common benefit that accrues to collaborative partners through combination, exchange, and co-development of idiosyncratic resources (Dyer and Singh 1998). It is joint competitive advantage and focuses on joint value creation in dyadic relationship. Supply chain partners work together toward common goals and achieve more mutual benefits than can be achieved by acting alone (Mentzer et al. 2001; Stank et al. 2001; Manthou et al. 2004; Sheu et al. 2006). In contrast, according to the extended resource bases view, competitive advantage focuses more on value appropriation by both appropriating relational rent (i.e., common benefits) and unilaterally accumulating spillover rents that produce private benefits (Lavie 2006).

Based on the ERBV (Lavie 2006), collaborative advantage can be understood as a function of the combined value and rarity of all shared resources among supply chain partners (i.e., relational rents) while competitive advantage of a firm depends on the total value and rarity of the firm's own shared and non-shared resources (i.e., internal rents) and resources interactions with partners (i.e., appropriated relational rents and spillover rents). When collaboration is formed, each partner endows a subset of its resources to the collaboration with the expectation of generating common benefits from the shared resources of both firms (Lavie 2006).

## 1.3  Gaps in Literature

Supply chain collaboration is not yet well investigated. Although many case studies, conceptual papers, and empirical research articles have been published (Buckley and Casson 1996; Mariti and Smiley 1996; Pfeffer and Nowak 1996; Kay 1997; Lee et al. 1997; Casson 1998; Dyer and Singh 1998; Tuten and Urban 2001; Lambert et al. 2004; Goffin et al. 2006; Allred et al. 2011; Nyaga et al. 2010; Cao and Zhang 2011; Fawcett et al. 2011), more needs to be done to better understand the concept of supply chain collaboration. Prior understanding of supply chain collaboration has been obscured by the implicit assumption that partnerships are always desirable (Boddy et al. 2000; Simatupang and Sridharan 2005a). Little attention has been paid to capturing the various characteristics that represent different aspects or areas of collaboration (Mentzer et al. 2000). The variety of conditions that affect or characterize supply chain collaboration is undervalued (Goffin et al. 2006). There are several gaps in the literature.

First, although the advantages of supply chain collaboration are widely acknowledged in the literature, the exact nature and attributes of supply chain collaboration are not well comprehended. Sheu et al. (2006) point out that the literature on supply chain collaboration is fragmented in that different disciplines often focus on only a small number of different factors. Research in marketing and management focuses on factors such as commitment (Handfield and Bechtel 2002), studies in operations management concentrate on factors such as information sharing and inventory systems (Srinivasan et al. 1994), and information systems researchers focus on IT capabilities (Grover et al. 2002). Fragmentation has inhibited the thorough understanding of phenomena (Barringer and Harrison 2000). Prior work fails to provide a comprehensive conceptualization of supply chain collaboration, which consequently limits our ability to explain and evaluate the level of collaborative efforts (Saeed 2004). Thus, a thorough understanding of the characteristics of supply chain collaboration is extremely important.

Second, in characterizing and conceptualizing supply chain collaboration, researchers focus more on process integration (e.g., goal congruence, decision synchronization, incentive alignment, and resource sharing) and less on collaborative communication and joint knowledge creation. Miscommunication, which causes conflicts and misunderstanding between supply chain partners, is recognized as the reason for many collaboration failures (Tuten and Urban 2001). Communication is the glue that holds supply chain partners together (Mohr and Nevin 1990). Further, collaborations between supply chain partners are not merely pure transactions, but long term partnerships which leverage information sharing and market knowledge creation for sustainable competitive advantage (Malhotra et al. 2005).

Third, in investigating IOS use to facilitate supply chain collaboration, prior studies focus on IOS enabled relationship-specific process integration between partners (Venkatraman and Zaheer 1990; Lee et al. 1997; Hart and Saunders 1998; Lambert et al. 2004; Saeed et al. 2005). Other roles of IOS, such as IOS use for communication to enhance supply chain partners' collaborative communication and IOS use for intelligence to improve supply chain partners' joint knowledge creation, have been largely unexplored in extant literature.

Fourth, in researching the antecedents or conditions that lead to or affect supply chain collaboration, prior studies focus on the use of IOS but simplify or ignore its culture context (Jagdev and Thoben 2001; Parks 2001; Skjoett-Larsen et al. 2003; Holweg et al. 2005; Gopal and Gosain 2010; Rai et al. 2012). Although IOS use is necessary for supply chain collaboration to succeed, organizational culture must be taken into consideration simultaneously (McCarter et al. 2005; Du et al. 2011). Many supply chain collaborations fail due to incompatible organizational culture and the complexities involved (Kanter 1989; Culpan 1993; Spekman et al. 1998).

Moreover, considerable difficulties exist among supply chain partners due to mutual distrust during collaboration (Simatupang and Sridharan 2002; Du et al. 2011). In the IOS enabled supply chain or virtual collaborative relationships, a high level of trust is required for collaboration to succeed (Ararwal and Shankar 2003; Gallivan and Depledge 2003; Paul and McDaniel 2004). Trust, as a critical determinant in establishing a relational mode of governance structure, is discounted in the current literature (Kumar et al. 1998). In spite of discussions about the need for trust in collaborative activities, there is a scarcity of large-scale empirical studies showing that trust actually has any impact on IOS enabled supply chain collaboration. Furthermore, there is a lack of accurate operationalization of trust and related concepts such as supply chain collaboration and performance outcomes, which hinders the empirical testing of their relationships.

Finally, in investigating the consequences of supply chain collaboration, existing literature ignores the collaborative advantage or joint competitive advantage achieved through collaboration.

In the extant literature, different perspectives have been taken in explaining supply chain collaboration. Some researchers use technical–economic perspectives such as transaction cost theory (Williamson 1975; Malone et al. 1987; Barringer and Harrison 2000; Kaufman et al. 2000; Croom 2001; Nesheim 2001; Son et al. 2005) and resource based theory (Barney 1991; Knudsen 2003; Park et al. 2004; Verwaal and Hesselmans 2004; Saeed et al. 2005). They argue that supply chain collaboration (1) reduces transaction costs; (2) requires asset-specific investments, which increase switching costs and lock-in partners; (3) is imperfectly imitable. Thus, collaboration can reduce uncertainty and opportunism and lead to process efficiency and competitive advantage.

Some scholars take socio-political perspectives, such as resource dependence theory (Kling 1980; Barringer and Harrison 2000) and social exchange theory (Blau 1964; Das and Teng 2002; Son et al. 2005; Thomas and Ranganathan 2005), to explain supply chain collaboration. They argue that there are many sources of resources that make some partners more powerful than others. The self-interested powerful firms take advantage of the less powerful partners by obtaining large portions of benefits, therefore leading to negotiation, conflicts, and politics, which further make collaboration very complex and eventually disintegrate supply chain collaboration.

Both technical–economic and socio-political perspectives seem useful to explain supply chain collaboration; however they do not capture the full picture of the phenomenon. Other complementary perspectives such as trust-based rationalism (Kumar et al. 1998) and knowledge based view (Nonaka and Takeuchi 1995; Barringer and Harrison 2000; Zahra and George 2002; Verwaal and Hesselmans 2004; Malhotra et al. 2005) also contribute to the comprehension of the concept. Trust based rationalism extends technical–economic theories by examining the non-contractual based reasons for participating in an exchange, e.g., embeddedness and trustworthiness, and gaining social capitals. It argues that supply chain collaboration is governed by implicit social contracts based on trust and social influence.

Learning and knowledge perspectives regard supply chain collaboration as partner-enabled market knowledge creation and value innovation process through rich information sharing and IOS use (Malhotra et al. 2005). Supply chain collaboration enables firms to enhance absorptive capacity by acquiring, assimilating, transforming, and exploiting real-time information between partners and further improve operational efficiency and knowledge creation. Supply chain collaboration is a living system where all partners grow together (Kanter 1994). By joint knowledge creation, firms gain intellectual capital and sustained collaborative advantage.

## 1.4 Research Questions

The objective of the study is to uncover the nature and characteristics, antecedents, and consequences of supply chain collaboration from multiple theoretical perspectives. To achieve this, the current study aims to shed light on the role of IOS use in supply chain collaboration by investigating the following research questions:

1 What is the nature of supply chain collaboration?

- To what extent do firms share information and integrate process with their supply chain partners?

- To what extent do firms communicate with their supply chain partners?
- To what extent do firms jointly create knowledge with their supply chain partners?

2. What factors differentiate successful from unsuccessful supply chain collaborations in their use of IOS?

   - What roles does IOS play in supply chain collaboration?
   - What roles does culture play in IOS enabled supply chain collaboration?
   - What roles does trust play in IOS enabled supply chain collaboration?

3. What benefits can firms obtain out of supply chain collaboration? Why do firms govern external transactions through relational and collaborative mechanisms rather than market mechanisms?

   - How should collaborative advantage be addressed?
   - How does supply chain collaboration affect firms' collaborative advantage and financial performance?
   - Does collaborative advantage completely mediate the relationship between supply chain collaboration and firm performance?
   - Does firm size moderate the relationships among supply chain collaboration, collaborative advantage, and firm performance?

## 1.5 Purported Contribution

The study contributes to the knowledge on IOS enabled supply chain collaboration by providing theoretical insights into and empirical findings on the above research questions. Through pooling an extensive set of factors from multiple perspectives, the research extends our understanding of the attributes of supply chain collaboration, the forces leading to the development of supply chain partnership, and issues involved in creating and managing the partnership. A better understanding of supply chain collaboration leads to better management of it. Specifically, the research intends to make the following contributions:

- Defining and conceptualizing supply chain collaboration by adding previously under explored components of collaborative communication and joint knowledge creation in addition to the widely studied foundation components of information sharing and process integration. Based on the rationale of co-creation of value, supply chain collaboration is conceptualized as having seven interconnecting elements (i.e. information sharing, goal congruence, decision synchronization, incentive alignment, resource sharing, collaborative communication, and joint knowledge creation) that are necessary and sufficient to define the occurrence of collaborative efforts. This comprehensive definition provides a way to explain supply chain collaboration more precisely.

- Proposing and empirically testing a theoretical framework that relates supply chain collaboration, its antecedents (IT resources, IOS appropriation, collaborative culture, and trust) and its consequences (collaborative advantage and firm performance). The framework is grounded in extensive literature and based on multiple perspectives (e.g., transaction cost economics, resource based view, social exchange theory, trust based rationalism, and knowledge and learning perspective).
- Defining IOS appropriation as patterns, modes, or fashion of IOS use and exploring its different roles (i.e., integration, communication, and intelligence) in supply chain collaboration. The role of culture and trust in IOS enabled supply chain collaboration is also investigated.
- Exploring the collaborative advantage of supply chain collaboration and its impact on firm performance.
- Developing reliable and valid instruments of key constructs to support research on supply chain collaboration. The instruments will also be useful for assessing the level of supply chain collaboration and identifying the best practice.

The rest of the book is organized as follows. Chapter 2 reviews the theoretical bases and proposes the research model. Chapter 3 discusses the antecedents of supply chain collaboration. Chapter 4 explores the nature and characteristics of supply chain collaboration. Chapter 5 focuses on the consequences of supply chain collaboration—collaborative advantage and firm performance. Chapter 6 describes structured interview and Q-sort. Chapter 7 presents large-scale analysis and hypotheses testing. Chapter 8 provides a discussion of research and managerial insights.

# References

Adegbesan, J. A., & Higgins, M. J. (2011). The intra-alliance division of value created through collaboration. *Strategic Management Journal, 32*(2), 187–211.

Allred, C. R., Fawcett, S. E., Wallin, C., & Magnan, G. M. (2011). A dynamic collaboration capability as a source of competitive advantage. *Decision Sciences, 42*(1), 129–161.

Ararwal, A., & Shankar, R. (2003). On-line trust building in e-enabled supply chain. *Supply Chain Management: An International Journal, 8*(4), 324–334.

Barney, J. (1991). Firm resources and sustained competitive advantage. *Journal of Management, 17*(1), 99–120.

Barratt, M. (2003). Positioning the role of collaborative planning in grocery supply chains. *The International Journal of Logistics Management, 14*(2), 53–66.

Barratt, M., & Oliveira, A. (2001). Exploring the experiences of collaborative planning initiatives. *International Journal of Physical Distribution and Logistics Management, 31*(4), 66–89.

Barringer, B. R., & Harrison, J. S. (2000). Walking a tightrope: Creating value through interorganizational relationships. *Journal of Management, 26*(3), 367–403.

Blau, P. M. (1964). *Exchange and power in social life*. New York: Wiley.

Blois, J. K. (1996). Vertical quasi-integration. In P. J. Buckley & J. Michie (Eds.), *Firms, organizations and contracts: A reader in industrial organization* (pp. 320–338). New York: Oxford University Press.

Boddy, D., Macbeth, D., & Wagner, B. (2000). Implementing collaboration between organizations: An empirical study of supply chain partnering. *Journal of Management Studies, 37*(7), 1003–1019.

Buckley, P. J., & Casson, M. (1996). Joint ventures. In P. J. Buckley & J. Michie (Eds.), *Firms, organizations and contracts: A reader in industrial organization* (pp. 410–428). New York: Oxford University Press.

Callioni, G., & Billington, C. (2001). Effective collaboration: Hewlett-Packard takes supply chain management to another level. *OR/MS Today, 28*(5), 34–39.

Cao, M., & Zhang, Q. (2011). Supply chain collaboration supply chain collaboration: Impact on collaborative advantage and firm performance. *Journal of Operations Management, 29*(3), 163–180.

Casson, M. (1998). *Information and organization: A new perspective on the theory of the firm.* New York: Oxford University Press.

Chen, I. J., & Paulraj, A. (2004). Towards a theory of supply chain management: The constructs and measurements. *Journal of Operations Management, 22*, 119–150.

Coff, R. (2010). The coevolution of rent appropriation and capability development. *Strategic Management Journal, 31*, 711–713.

Croom, S. (2001). Restructuring supply chains through information channel innovation. *International Journal of Operations and Production Management, 21*(4), 504–515.

Croom, S., Romano, P., & Giannakis, M. (2000). Supply chain management: An analytical framework for critical literature review. *European Journal of Purchasing and Supply Management, 6*(1), 67–83.

Crum, C., & Palmatier, G. E. (2004). Demand collaboration: What's holding us back? *Supply Chain Management Review, 8*(1), 54–61.

Culpan, R. (1993). Multinational competition and cooperation: Theory and practice. In R. Culpan (Ed.), *Multinational strategic alliances.* New York: Haworth Press, Inc.

Das, T. K., & Teng, B. S. (2002). Social exchange theory of strategic alliances. In F. J. Contractor & P. Lorange (Eds.), *Cooperative strategies and alliances* (pp. 439–460). Oxford, UK: Elsevier Science.

Dell, M., & Fredman, C. (1999). *Direct from Dell: Strategies that revolutionized an industry.* London: Harper Collins Business.

Doz, Y. L., & Hamel, G. (1998). *Alliance advantage.* Boston: Harvard Business School Press.

Du, S., Bhattacharya, C., & Sankar, S. (2011). Corporate social responsibility and competitive advantage: Overcoming the trust barrier. *Management Science, 57*(9), 1528–1545.

Dyer, J. H. (2000). *Collaborative advantage: Winning through extended enterprise supplier networks.* New York, NY: Oxford University Press.

Dyer, J. H., & Singh, H. (1998). The relational view: Cooperative strategy and sources of interorganizational competitive advantage. *Academy of Management Review, 23*(4), 660–679.

Eisenhardt, K., & Schoonhoven, C. B. (1996). Resource-based view of strategic alliance formation in entrepreneurial firms: Strategic needs and social opportunities for cooperation. *Organization Science, 7*(2), 136–150.

Fawcett, S. E., & Magnan, G. M. (2004). Ten guiding principles for high-impact SCM. *Business Horizon, 47*(5), 67–74.

Fawcett, S. E., Wallin, C., Allred, C., Fawcett, A., & Magnan, G. M. (2011). Information technology as an enabler of supply chain collaboration: A dynamic-capabilities perspective. *Journal of Supply Chain Management, 47*(1), 38–59.

Fawcett, S. E., Fawcett, A., Watson, B., & Magnan, G. (2012). Peeking inside the black box: toward an understanding of supply chain collaboration dynamics. *Journal of Supply Chain Management, 48*(1), 44–72.

Flynn, B. B., Huo, B., & Zhao, X. (2010). The impact of supply chain integration on performance: A contingency and configuration approach. *Journal of Operations Management, 28*(1), 58–71.

Gallivan, J. M., & Depledge, G. (2003). Trust, control and the role of interorganizational systems in electronic partnerships. *Information Systems Journal, 13*, 159–190.

Goffin, K., Lemke, F., & Szwejczewski, M. (2006). An exploratory study of close supplier-manufacturer relationships. *Journal of Operations Management, 24*(2), 189–209.

Gopal, A., & Gosain, S. (2010). The role of organizational controls and boundary spanning in software development outsourcing: implications for project performance. *Information Systems Research, 21*(4), 960–982.

Grover, V., & Kohli, R. (2012). Cocreating IT value: New capabilities and metrics for multifirm environments. *MIS Quarterly, 36*(1), 225–232.

Grover, V., Teng, J., & Fiedler, K. (2002). Investigating the role of information technology in building buyer-supplier relationships. *Journal of Association for Information Systems, 3*, 217–245.

Gulati, R. (1995). Social structure and alliance formation: A longitudinal analysis. *Administrative Science Quarterly, 40*(4), 619–652.

Handfield, R. B., & Bechtel, C. (2002). The role of trust and relationship structure in improving supply chain responsiveness. *Industrial Marketing Management, 31*(4), 367–382.

Handfield, R. B., & Nichols, E. L, Jr. (2002). *Supply chain redesign: Transforming supply chains into integrated value systems. Financial Times*. London: Prentice-Hall.

Hart, P., & Saunders, C. (1998). Emerging electronic partnerships: Antecedents and dimensions of EDI use from the supplier's perspective. *Journal of Management Information Systems, 14*(4), 87–111.

Hatfield, L., & Pearce, J. A, I. I. (1994). Goal achievement and satisfaction of joint venture partners. *Journal of Business Venturing, 9*(5), 423–449.

Henderson, R., & Cockburn, I. (1994). Measuring competence? Exploring firm effects in pharmaceutical research. *Strategic Management Journal, 15*(8), 63–84.

Holland, C. P. (1995). Cooperative supply chain management: The impact of interorganizational information systems. *Journal of Strategic Information Systems, 4*(2), 117–133.

Holweg, M., Disney, S., Holmström, J., & Småros, J. (2005). Supply chain collaboration: Making sense of the strategy continuum. *European Management Journal, 23*(2), 170–181.

Jagdev, H. S., & Thoben, K. D. (2001). Anatomy of enterprise collaboration. *Production Planning and Control, 12*(5), 437–451.

Jap, S. D. (1999). Pie-expansion efforts: Collaboration processes in buyer-supplier relationships. *Journal of Marketing Research, 36*(4), 461–476.

Johnson, J. J., & Sohi, R. S. (2003). The development of interfirm partnering competence: Platforms for learning, learning activities and consequences of learning. *Journal of Business Research, 56*(9), 757–766.

Kalwani, M. U., & Narayandas, N. (1995). Long-term manufacturer–supplier relationships: Do they pay? *Journal of Marketing, 59*(1), 1–15.

Kanter, R. M. (1989). *When giants learn to dance*. New York: Simon and Schuster.

Kanter, R. M. (1994). Collaborative advantage: The art of alliances. *Harvard Business Review, 72*, 96–108.

Kaufman, A., Wood, C. H., & Theyel, G. (2000). Collaboration and technology linkages: A strategic supplier typology. *Strategic Management Journal, 21*(6), 649–663.

Kay, N. M. (1997). *Patterns in Corporate Evolution*. New York: Oxford University Press.

Khanna, T., Gulati, R., & Nohria, N. (1998). The dynamics of learning alliances: Competition, cooperation, and relative scope. *Strategic Management Journal, 19*, 193–210.

Kling, R. (1980). Social analysis of computing: Theoretical perspectives in recent empirical research. *ACM Computing Surveys, 12*(1), 61–110.

Knudsen, D. (2003). Aligning corporate strategy, procurement strategy and e-procurement tools. *International Journal of Physical Distribution and Logistics Management, 38*(8), 720–734.

Kogut, B. (1988). Joint ventures: Theoretical and empirical perspectives. *Strategic Management Journal, 9*(4), 319–332.

Koka, B. R., & Prescott, J. E. (2002). Strategic alliances as social capital: A multidimensional view. *Strategic Management Journal, 23*(9), 795–816.

Konsynski, B. R., & McFarlan, E. W. (1990). Information partnership–Shared data, shared scale. *Harvard Business Review, 68*(5), 114–121.

Kumar, M. V. (2011). Are joint ventures positive sum games? The relative effects of cooperative and noncooperative behavior. *Strategic Management Journal, 32*(1), 32–54.

Kumar, K., Van Dissel, H. G., & Bielli, P. (1998). The merchant of Prato revisited: Toward a third rationality of information systems. *MIS Quarterly, 22*(2), 199–226.

Lambert, D. M., & Cooper, M. C. (2000). Issues in supply chain management. *Industrial Marketing Management, 29*(1), 65–83.

Lambert, D. M., Cooper, M. C., & Pugh, J. D. (1998). Supply chain management: Implementation issues and research opportunities. *International Journal of Logistics Management, 9*(2), 1–19.

Lambert, D. M., Emmelhainz, M. A., & Gardner, J. T. (2004). Supply chain partnerships: Model validation and implementation. *Journal of Business Logistics, 25*(2), 21–42.

Lavie, D. (2006). The competitive advantage of interconnected firms: An extension of the resource-based view. *Academy of Management Review, 31*(3), 638–658.

Lee, H.L., & Whang, S. (2001). E-Business and supply chain integration. *Stanford Global Supply Chain Management Forum,* SGSCMF-W2-2001.

Lee, H. L., Padmanabhan, V., & Whang, S. (1997). The bullwhip effect in supply chain. *Sloan Management Review, 38*(3), 93–102.

Lejeune, N., & Yakova, N. (2005). On characterizing the 4 C's in supply China management. *Journal of Operations Management, 23*(1), 81–100.

Liker, J. K., & Choi, T. Y. (2005). Building deep supplier relationships. *Harvard Business Review, 83*(1), 104–113.

Malhotra, A., Gasain, S., & El Sawy, O. A. (2005). Absorptive capacity configurations in supply chains: Gearing for partner-enabled market knowledge creation. *MIS Quarterly, 29*(1), 145–187.

Malone, T. W., Yates, J., & Benjamin, R. I. (1987). Electronic markets and electronic hierarchies. *Communications of the ACM, 30*(6), 484–497.

Manthou, V., Vlachopoulou, M., & Folinas, D. (2004). Virtual e-Chain (VeC) model for supply chain collaboration. *International Journal of Production Economics, 87*(3), 241–250.

Mariti, P., & Smiley, R. H. (1996). Co-operative agreements and the organization of industry. In P. J. Buckley & J. Michie (Eds.), *Firms, organizations and contracts: A reader in industrial organization* (pp. 276–292). New York: Oxford University Press.

McCarter, M. W., Fawcett, S. E., & Magnan, G. M. (2005). The effect of people on the supply chain world: Some overlooked issues. *Human Systems Management, 24*(3), 197–208.

Mentzer, J. T., Foggin, J. H., & Golicic, S. L. (2000a). Collaboration: The enablers, impediments, and benefits. *Supply Chain Management Review, 5*(6), 52–58.

Mentzer, J. T., Min, S., & Zacharia, Z. G. (2000b). The nature of interfirm partnering in supply chain management. *Journal of Retailing, 76*(4), 549–568.

Mentzer, J. T., DeWitt, W., Keebler, J. S., Min, S., Nix, N. W., Smith, C. D., et al. (2001). Defining supply chain management. *Journal of Business Logistics, 22*(2), 1–25.

Min, S., Roath, A., Daugherty, P. J., Genchev, S. E., Chen, H., & Arndt, A. D. (2005). Supply chain collaboration: What's happening? *International Journal of Logistics Management, 16*(2), 237–256.

Mohr, J., & Nevin, J. R. (1990). Communication strategies in marketing channels: A theoretical perspective. *Journal of Marketing, 54*(4), 36–51.

Mohr, J., & Spekman, R. E. (1994). Characteristics of partnership success: partnership attributes, communication behavior, and conflict resolution techniques. *Strategic Management Journal, 15*(2), 135–152.

Narus, J. A., & Anderson, J. C. (1996). Rethinking distribution: Adaptive channels. *Harvard Business Review, 74*(4), 112–120.

Nesheim, T. (2001). Externalization of the core: Antecedents of collaborative relationships with suppliers. *European Journal of Purchasing and Supply Management, 7*(4), 217–225.

Newman, R. G. (1988). Single source qualification. *Journal of Purchasing and Materials Management, 24*(2), 10–16.

Niederkofler, M. (1991). The evolution of strategic alliances: Opportunities for managerial influence. *Journal of Business Venturing, 6*(4), 237–257.

Nonaka, I., & Takeuchi, H. (1995). *The knowledge creating company.* New York: Oxford University Press.

Nyaga, G., Whipple, J., & Lynch, D. (2010). Examining supply chain relationships: Do buyer and supplier perspectives on collaborative relationships differ? *Journal of Operations Management, 28*(2), 101–114.

Park, N. K., Mezias, J. M., & Song, J. (2004). A resource-based view of strategic alliances and firm value in the electronic marketplace. *Journal of Management, 30*(1), 7–27.

Parks, L. (1999). CRP investment pays off in many ways. *Drug Store News, 21*(2), 26.

Parks, L. (2001). Wal-Mart gets onboard early with collaborative planning. *Drug Store News, 23*(2), 14.

Paul, D. L., & McDaniel, R. R, Jr. (2004). A field study of the effect of interpersonal trust on virtual collaborative relationship performance. *MIS Quarterly, 28*(2), 183–227.

Pfeffer, J., & Nowak, P. (1996). Joint ventures and interorganizational interdependence. In P. J. Buckley & J. Michie (Eds.), *Firms, organizations and contracts: A reader in industrial organization* (pp. 385–409). New York: Oxford University Press.

Porter, M. E. (1985). Technology and competitive advantage. *Journal of Business Strategy, 5*(3), 60–78.

Powell, W. W., Kogut, K. W., & Smith-Doerr, L. (1996). Interorganizational collaboration and the locus of innovation: Networks of learning in biotechnology. *Administrative Science Quarterly, 41*(1), 116–145.

Prahalad, C.K., & Ramaswamy, V. (2001). The collaboration continuum. *Optimize Magazine,* November, 31–39.

Rai, A., Pavlou, P., Im, G., & Du, S. (2012). Interfirm IT capability profiles and communications for cocreating relational value: evidence from the logistics industry. *MIS Quarterly, 36*(1), 233–267.

Sabath, R. E., & Fontanella, J. (2002). The unfulfilled promise of supply chain collaboration. *Supply Chain Management Review, 6*(4), 24–29.

Saeed, K.A. (2004). Information technology antecedents to supply chain integration and firm performance. *Unpublished Dissertation.* University of South Carolina.

Saeed, K. A., Malhotra, M. K., & Grover, V. (2005). Examining the impact of interorganizational systems on process efficiency and sourcing leverage in buyer–supplier dyads. *Decision Sciences, 36*(3), 365–396.

Sanders, N. R., & Premus, R. (2005). Modeling the relationship between IT capability, collaboration, and performance. *Journal of Business Logistics, 26*(1), 1–24.

Sheu, C., Yen, H. R., & Chae, D. (2006). Determinants of supplier-retailer collaboration: Evidence from an international study. *International Journal of Operations and Production Management, 26*(1), 24–49.

Simatupang, T. M., & Sridharan, R. (2002). The collaborative supply chain. *International Journal of Logistics Management, 13*(1), 15–30.

Simatupang, T. M., & Sridharan, R. (2005). An Integrative framework for supply chain collaboration. *International Journal of Logistics Management, 16*(2), 257–274.

Skjoett-Larsen, T., Thernoe, C., & Andersen, C. (2003). Supply chain collaboration. *International Journal of Physical Distribution and Logistics Management, 33*(6), 531–549.

Snyder, L., & Shen, Z. (2011). *Fundamentals of supply chain theory.* New York: Wiley.

Son, J., Narasimhan, S., & Riggins, F. J. (2005). Effects of relational factors and channel climate on EDI usage in the customer-supplier relationship. *Journal of Management Information Systems, 22*(1), 321–353.

Spekman, R. E. (1988). Strategic supplier selection: Understanding long-term buyer relationships. *Business Horizons, 31*(4), 75–81.

Spekman, R. E., Forbes, T. M., Isabella, L. A., & MacAvoy, T. C. (1998). Alliance management: A view from the past and look to the future. *Journal of Management Studies, 35*(6), 747–772.

Srinivasan, K., Kekre, S., & Mukhopadhyay, T. (1994). Impact of electronic data interchange technology on JIT shipments. *Management Science, 40*(10), 1291–1304.

Stank, T. P., Keller, S. B., & Daugherty, P. J. (2001). Supply chain collaboration and logistical service performance. *Journal of Business Logistics, 22*(1), 29–48.

Stuart, F. I., & McCutcheon, D. (1996). Sustaining strategic supplier alliances. *International Journal of Operation and Production Management, 16*(10), 5–22.

Teo, T., Nishant, R., Goh, M., & Agarwal, S. (2011). Leveraging collaborative technologies to build a knowledge sharing culture at HP Analytics. *MIS Quarterly Executive, 10*(1), 1–18.

Thomas, D., & Ranganathan, C. (2005). Enabling e-business transformation through alliances: Integrating social exchange and institutional perspectives. *Proceedings of the 38th Hawaii International Conference on System Sciences.*

Thompson, J. D. (1967). *Organizations in action: Social science bases of administrative theory.* New York: McGraw-Hill book Company.

Tuten, T. L., & Urban, D. J. (2001). An expanded model of business-to-business partnership foundation and success. *Industrial Marketing Management, 30*(2), 149–164.

Venkatraman, N., & Zaheer, A. (1990). Electronic integration and strategic advantage: A quasi-experimental study in the insurance industry. *Information Systems Research, 1*(4), 377–393.

Verdecho, M., Alfaro-Saiz, J., Rodriguez–Rodriguez, R., & Ortiz-Bas, A. (2012). A multi-criteria approach for managing inter-enterprise collaborative relationships. *Omega, 40*(3), 249–263.

Verwaal, E., & Hesselmans, M. (2004). Drivers of supply network governance: An explorative study of the dutch chemical industry. *European Management Journal, 22*(4), 442–451.

Wagner, S., Eggert, A., & Lindemann, E. (2010). Creating and appropriating value in collaborative relationshlps. *Journal of Business Research, 63*, 840–848.

Webster's Third New International Dictionary. 1966. William Benton, Chicago.

Williamson, O. E. (1975). *Markets and hierarchies.* Englewood Cliffs, NJ: Prentice-Hall.

Williamson, O. E. (1985). *The economic institutions of capitalism.* New York, NY: The Free Press.

Zahra, S. A., & George, G. (2002). The net-enabled business innovation cycle and the evolution of dynamic capabilities. *Information Systems Research, 13*(2), 147–150.

# Chapter 2
# Theory and Theoretical Framework

**Abstract** In the extant literature, different perspectives and theories have been taken in explaining supply chain collaboration. In this chapter, we examine supply chain collaboration using the following ten theories: uncertainty reduction theory, transaction cost economics, resource based view, relational view, extended resource based view, resource dependence theory, social exchange theory, social dilemma theory, trust based rationalism, and learning and knowledge perspective. We describe and compare the relative strength and weakness of each theory in situating the phenomenon of supply chain collaboration. These multiple perspectives provide us with insights into the nature, forms, contents, and forces of supply chain collaboration. We also draw on the key concepts from theories and literature and use them to propose and develop the theoretical framework where supply chain collaboration is the central concept.

## 2.1 Uncertainty Reduction Perspective

The theoretical literature on supply chain collaboration is diversified representing multiple perspectives. The diverse literature reflects the versatile nature of supply chain collaboration involving a variety of motives and objectives (Barringer and Harrison 2000; Hitt 2011; Verdecho et al. 2012; Fawcett et al. 2012). This study examines supply chain collaboration from multiple perspectives: (1) technical-economic perspective, e.g. uncertainty reduction, transaction cost economics, resource based view, relational view, and extended resource based view; (2) socio-political perspective, e.g. resource dependence theory, social exchange theory, and social dilemma theory; (3) trust based rationalism; and (4) learning and knowledge perspective. These multiple perspectives provide us with insights into the nature, forms, contents, and forces of supply chain collaboration.

M. Cao and Q. Zhang, *Supply Chain Collaboration*,
DOI: 10.1007/978-1-4471-4591-2_2, © Springer-Verlag London 2013

Uncertainty has long been viewed as a dominant contingency and is one of the underlying determinants of high transaction costs (Williamson 1975). Reducing uncertainty via transparent information flow is a key objective in supply chain collaboration (Holweg et al. 2005). Market and technological uncertainty can effectively be dealt with through partnerships where supply chain partners share information of unexpected events and developments (Verwaal and Hesselmans 2004). The intense communication between supply chain partners also reduces behavioral uncertainty (e.g., opportunism) (Wuyts and Geyskens 2005). If information is not shared between partners, non-transparent demand patterns will cause demand amplification and bullwhip effect. This leads to poor service levels, high inventories, and frequent stock-outs (Lee et al. 1997). Thus, when facing uncertainty, firms will tend to collaborate with partners in building long-term relationship.

## 2.2 Transaction Cost Economics

Transaction cost economics (TCE) is one of the most influential theories on IOS use and interfirm collaboration (Williamson 1975; Barringer and Harrison 2000; Nesheim 2001; Wever et al. 2012). TCE suggests that a firm organize its cross-organizational activities to minimize production costs within the firm and transaction costs within markets. According to TCE, the decision to use either vertical integration or market mechanisms depends on the relative monitoring costs that arise from bounded rationality and uncertainties due to partners' self-interest and opportunism (Kaufman et al. 2000). TCE thinks that IOS use can reduce transaction costs (e.g., monitoring costs) by specific asset investments, which diminish opportunistic behaviors (Son et al. 2005).

Williamson (1975) identifies markets and hierarchies as two modes of organizing. Collaboration emerges as the third alternative. Supply chain collaboration helps prevent the problems arising from both markets and hierarchies (Koh and Venkatraman 1991). It helps firms reduce the opportunism and monitoring costs that are inbuilt in market transactions through process integration and mutual trust, thus reduce the probability that partners behave opportunistically (Kaufman et al. 2000; Croom 2001). Supply chain collaboration also helps firms avoid internalizing an activity that they do not excel at (Harrigan 1988).

In spite of TCE's usefulness, many scholars notice its limitation. TCE is restricted to the efficiency rationale for supply chain collaboration. Supply chain collaboration may form for other reasons such as knowledge creation. In addition, organizational contexts (e.g. culture, power, dependence, and trust) that may affect collaborative efforts are assumed away (Barringer and Harrison 2000; Duffy and Fearne 2004). In reality, few supply chain collaborations are purely based on the consideration of transaction costs (Faulkner 1995).

## 2.3 Resource Based View

Resource based view (RBV) receives much attention in explaining supply chain collaboration. The key concepts of RBV are resources, capabilities, and strategic assets (Barney 1991). RBV argues that variance in firm performance can be explained by strategic resources, such as core competence (Prahalad and Hamel 1990), dynamic capability (Amit and Schoemaker 1993; Teece et al. 1997), and absorptive capacity (Cohen and Levinthal 1990). Firms that combine resources in a unique way may achieve an advantage over their competing firms who are unable to do so (Dyer and Singh 1998). By owning scarce resources and assets and excelling in core competencies and capabilities, firms can reach a market advantage and gain a sustained competitive advantage (Knudsen 2003). RBV claims that electronic integration by specific asset investments enables partnering firms to build competitive advantage because of their rare, valuable, non-substitutable, and difficult-to-imitate nature (Barney 1991; Knudsen 2003).

Resource complementarity or the need for particular resources is another reason for supply chain collaboration (Knudsen 2003). By investments in relation-specific assets, substantial knowledge exchange, combining complementary and scarce resources or capabilities, supply chain collaboration can create unique products, services or technologies (Knudsen 2003). Rents are generated through synergistic combination of assets, knowledge, or capabilities (Das and Teng 2000). The embeddedness of partnering firms' relational assets and the causal ambiguity are difficult for their competitors to copy (Hansen et al. 1997; Lorenzoni and Lipparini 1999; Jap 2001). Supply chain collaboration also enables firms to concentrate on their core competencies, which increase firm specific skills and realize economies of scale and learning effects, thereby improving their competitive positions (Barney 1991; Park et al. 2004; Verwaal and Hesselmans 2004).

## 2.4 Relational View

The relational view (RV) complements the RBV by arguing that critical resources may span firm boundaries (Dyer and Singh 1998). Firms can earn not only internal rents (i.e., Ricardian rents from scarcity of resources and quasi-rents from added value) but also relational rents. A relational rent is defined as a supernormal profit jointly generated in an exchange relationship that cannot be created by either firm in isolation and can only be created through the joint contributions of the collaborative partners (Dyer and Singh 1998; Lavie 2006). Relational rents are possible when collaborative partners combine and exchange idiosyncratic assets, knowledge, and capabilities through relation-specific investments, interfirm knowledge-sharing routines, complementary resource endowments, and effective governance mechanisms.

Collaborative advantage is based on the relational view, which elaborates on the mechanisms of joint value creation (i.e., interfirm rent generation). It argues relational rents accrue at the collaboration level for mutual benefits. Unlike studies that acknowledge the role of both private and common benefits (Hamel 1991; Khanna et al. 1998), the relational view emphasizes common benefits that collaborative partners cannot generate independently.

## 2.5  Extended Resource-Based View

Conventional RBV assumes firms must own or fully control the resources to create value. In the extended resource-based view (ERBV), resource accessibility, the right to employ resources or enjoy their associated benefits, enables firms to achieve advantages. Lavie (2006) extends the RBV by explaining how interconnected firms in dyadic collaboration/alliance combine external resources and internal resource endowments to achieve competitive advantage for the focal firm. According to Lavie (2006), the competitive advantage of a focal firm participating in an alliance/collaboration includes four elements: (1) internal rent (2) appropriated relational rent (3) inbound spillover rent, and (4) outbound spillover rent. Internal rent can be extracted from the focal firm's own shared and nonshared resources. Appropriated relational rent can be extracted only from the shared resources of both partners. Inbound spillover rent is the rent generated from the partner's shared and nonshared resources through knowledge leakage, inter-firm learning, relative absorptive capacity, and internalization of the partner's practices, whereas outbound spillover rent results from the transfer of benefits from the focal firm to the partner. The combination of internal rent, inbound spillover rent, and outbound spillover rent forms private benefits for the focal firm. Its competitive advantage depends on its private benefits and appropriated relational rent (i.e., appropriated common benefits).

In contrast, collaborative advantage is joint competitive advantage and come from a relational rent, a common benefit that accrues to collaborative partners (Dyer and Singh 1998). This type of rent cannot be generated individually by either collaborative partner. In addition, Lavie (2006) model extends prior research on joint value creation in dyadic alliance by considering unilateral accumulation of spillover rents that produce private benefits.

## 2.6  Resource Dependence Theory

Resource dependence theory (RDT) argues that firms must exchange with their environments to gain resources (Scott 1987). It centers solely on resources that must be acquired from external sources for a firm to survive or thrive (Barringer and Harrison 2000). The need for external resources makes firms depend on others.

To successfully manage dependencies, RDT argues that firms must gain control over vital resources to reduce reliance on others and increase others' reliance on them. It means firms should try to increase their power in their environments (Pfeffer and Nowak 1996; Thorelli 1986; Barringer and Harrison 2000). Supply chain collaboration provides such a way to helping firms to reach these goals.

Extending the logic of resource dependence theory from the firm level to the supply chain level, supply chain partners as a whole are less relying on their environments through resources sharing. Firms collaborate with their supply chain partners to acquire vital resources and to increase their power relative to other supply chains. However, the power may be unbalanced between partners because of different ownership of resources. This unbalance of power may create conflicts between partners if not well managed. Min et al. (2005) suggest the powerful firm in the supply chain should meet the less powerful partner's needs in mutually beneficial arrangements to strengthen the competitive power of the supply chain as a whole. Based on RDT, IOS are the instruments that, by easily accessing partners' resources, increase the supply chain's power over other firms or chains.

While RDT has its merits, it has limitations in explaining supply chain collaboration. RDT just argues that firms have to exchange with their environments to acquire necessary resources since no firm is self-contained. Transaction costs, competence development, and learning opportunities are not taken into consideration (Barringer and Harrison 2000).

## 2.7 Social Exchange Theory

Social exchange theory (SET) extends the technical–economic perspective by examining the non-contractual based reasons for participating in an exchange (Blau 1964; Das and Teng 2002; Thomas and Ranganathan 2005). Social exchanges differ from economic exchanges in that the specific benefits of exchange are not contractually and explicitly fully specified; partners have a social bond out of social influence. Supply chain collaboration can be explained by SET with the examination of social influence (e.g., power). According to SET, power is regarded as the most important sociological aspect of an interorganizational relationship when one firm needs to influence another's decisions. The exercise of power is often referred to as influence strategies (Son et al. 2005). These influences typically involve threats, punishment, rewards, and assistance.

## 2.8 Social Dilemma Theory

Social dilemma theory depicts situations where an individual's rational behavior causes suboptimal outcomes from the collective viewpoint (Dawes 1980; Kollock 1998). Research on social dilemma theory has concentrated on three social motives:

(1) individualism/independence—maximizing own outcomes regardless of others; (2) competition—maximizing own outcomes relative to others; and (3) cooperation—maximizing joint outcomes. The group members' goals—cooperation, competition, and independence—determine the interaction patterns of the group members, which in turn determine the group outcomes. Social dilemma theory concentrates on the tension between cooperation and competition. It conjectures that a member in the group can get higher benefits by defecting because he can obtain a larger piece of the pie than the portion that he deserves to have (Dawes 1980). It is confirmed that a self-interest member of an alliance is more likely to defect than cooperate (Zeng and Chen 2003). In contrast, if all members cooperate, the pie grows larger and accordingly each member gains higher although the percentage accrued by each member might not change (Dawes 1980). Social dilemma theory emphasizes that a member's short-term decision can lead to the potential long-term failure of the alliance.

## 2.9  Trust Based Rationalism

Trust based rationalism (TBR) employs a behavioral assumption of trustworthiness, fair play, responsibility, and altruism instead of betrayal, self-interest, and opportunism. It focuses on collaboration and cooperation rather than politics and conflicts as the primary interaction modes. Trust, relationship, and social capital are the key concepts in TBR. Trust is viewed as a critical determinant in establishing a relational mode of governance structure (Kumar et al. 1998). Continuing supply chain collaboration is based more on trust and equity than on monitoring and control capabilities (Kim et al. 2005).

Social capitals and relationships between partners arise from the foundation of trust. Trust reduces transaction costs and even eliminates the need for detailed contracts and governance mechanisms (Bromiley and Cummings 1995). While opportunism may create short-term benefits, it incurs costs in the long run because it lacks of reputation and trust (Kumar et al. 1998). Trust helps supply chain partners create a win–win strategy for collaborative advantage (Kumar and Van Dissel 1996).

## 2.10  Learning and Knowledge Perspective

Another rationale for explaining supply chain collaboration is that firms establish partnerships to exploit opportunities for knowledge creation and organizational learning (Kogut 1988; Hamel 1991; Mowery et al. 1996; Malhotra et al. 2005). Through knowledge creation and organizational learning, firms strengthen their competitive positions (Simonin 1997; Verwaal and Hesselmans 2004). In the face of high environmental uncertainty, it is important to have access to a broad and deep knowledge base in order to respond quickly to changing circumstances (Volberda 1998). Since great diversity of knowledge is distributed across the supply chain, collaboration provides an ideal platform for learning (Verwaal and Hesselmans 2004) and facilitates partner-enabled market knowledge creation (Malhotra et al. 2005).

Learning that takes place in supply chain collaboration can be divided into two kinds of activities: exploration and exploitation (March 1995; Barringer and Harrison 2000; Subramani 2004). Exploitation is to improve existing capabilities while exploration is to discover new opportunities (e.g., improve absorptive capacity) (Cohen and Levinthal 1990; Lane and Lubatkin 1998; Subramani 2004). How much a firm can learn through supply chain collaboration is determined by the firm's absorptive capacity, "the ability to recognize the value of new, external knowledge, assimilate it, and apply it to commercial ends." (Cohen and Levinthal 1990, p. 128). A firm's ability to learn is based on the employee quality, knowledge base, organizational culture, and the quality of IT systems (Kumar and Nti 1998).

Supply chain collaboration can also be an effective means of transferring knowledge and new technical skills across organizations. A firm may find it difficult to buy a particular skill in the marketplace because of its tacit nature (Mowery et al. 1996). It may acquire new skills and competencies by collaborating with firms that excel in that area (Barringer and Harrison 2000). However, the level of privileged information sharing needed for collaboration, in fear of risky information leakage, is not adequately addressed by the learning and knowledge theory.

## 2.11 Theoretical Framework

Each of the theories discussed above is useful but insufficient to capture the complexity involved in supply chain collaboration. By blending multiple theoretical perspectives, a more comprehensive picture of supply chain collaboration can be captured. In studying supply chain collaboration, a technical–economic view focuses on how IOS affects control and cost structures within the firm (i.e., production costs) and within markets (i.e., transaction costs) (Williamson 1975; Son et al. 2005). A socio-political perspective centers on how IOS and organizations interact while simultaneously taking organizational context (e.g. politics, power, conflicts, and culture) into consideration (Kling 1980; Barringer and Harrison 2000). Based on a behavioral assumption of trustworthiness rather than opportunism, trust based rationalism concentrates on trust, equity, and embeddedness rather than power and politics as the primary interaction mode in supply chain collaboration (Uzzi 1997; Kumar et al. 1998). A learning and knowledge perspective regards supply chain collaboration as partner-enabled market knowledge creation and value innovation process via IOS use (Malhotra et al. 2005).

Based on literature, supply chain collaboration consists of information sharing (Manthou et al. 2004) and process integration, such as goal congruence (Angeles and Nath 2001), joint decision making (Stank et al. 2001), joint planning (Mohr and Spekman 1994; Manthou et al. 2004), joint problem solving (Spekman et al. 1998; Stank et al. 2001), resource sharing (Sheu et al. 2006), and incentive alignment (Simatupang and Sridharan 2005), among independent supply chain partners (Stank et al. 1999; Sabath and Fontanella 2002, 1999; Simatupang and Sridharan 2002; Sheu et al. 2006). Over the past decades, firms have used IOS to develop collaborative relationships with their partners in the supply chain (Ragatz et al. 1997; Grover et al. 2002; Teo et al. 2003; Subramani 2004; Bagchi and Skjoett-Larsen 2005). Being integrated through shared information and process alignment, supply chain partners work as if they were a part of a single enterprise (Lambert and Christopher 2000).

While researchers have addressed some aspects of supply chain collaboration, they do not adequately highlight the need for collaborative communication as a critical partnership variable (Macneil 1980). Bleeke and Ernst (1993), p. xvi) argue: "The most carefully designed relationship will crumble without good, frequent communication." Communication difficulty is a prime cause of supply chain collaboration problems. Many problems in dealer channels can be resolved by developing appropriate strategies for communication (Mohr and Nevin 1990). "As the glue that holds together a channel of distribution" (Mohr and Nevin 1990, p. 36), communication is vital to the ongoing agreement of relationships (Grabner and Rosenberg 1969) and is the most important element to successful inter-firm exchange (Mohr et al. 1996).

Another overlooked but crucial variable in supply chain collaboration is joint knowledge creation. Supply chain collaboration should involve active generation and development of knowledge for retrieval and application in managing current and future business. Joint knowledge creation involves information acquisition, information dissemination, and shared interpretation of information (Slater and Narver 1995; Johnson and Sohi 2003). At the supply chain level, it is increasingly recognized that innovation involves learning in concert with partners (Harland et al. 2004) or collective entrepreneurship (Lundvall 1992). Both suppliers and customers are important sources of innovation (Von Hippel 1988; Nesheim 2001).

The study draws on the key concepts from theories and literature on information systems, supply chain management, operations management, marketing, and strategy, and uses them to situate and elaborate the theoretical model where supply chain collaboration is the central concept. As illustrated in Fig. 2.1, the framework provides a nomological network that describes the causal relationships among IT resources, IOS appropriation, collaborative culture, trust, supply chain collaboration, collaborative advantage, and firm performance. It can be used to study supply chain collaboration from a focal firm's perspective and test the hypotheses and structural relationships among the constructs.

The core construct of supply chain collaboration as co-creation of value consists of seven components: information sharing, goal congruence, decision synchronization, incentive alignment, resources sharing, collaborative communication, and

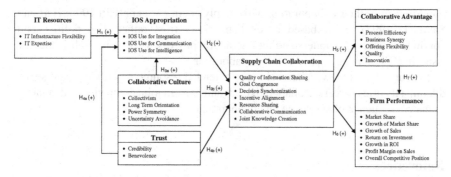

**Fig. 2.1** A framework for IOS-enabled supply chain collaboration

joint knowledge creation. These seven components add values to supply chain collaboration by either reducing costs and response time, or leveraging resources, or improving innovation. Information sharing is the fundamental component; all other components are the natural extension of it. Information sharing and process integration components (i.e., goal congruence, decision synchronization, incentive alignment, resource sharing) are considered as mechanisms to reduce costs based on transaction cost economics. Collaborative communication as an indispensable variable in supply chain collaboration is largely overlooked in the existing literature. Collaborative communication can reduce conflicts and improve relationships between partners. From the learning and knowledge perspective, joint knowledge creation is a key attribute of supply chain collaboration to enhance innovation and consolidate resources.

Based on transaction cost economics and resource based view, IT resources and IOS appropriation are powerful forces to enable supply chain collaboration. The existing literature does not distinguish between different roles of IOS use in supply chain collaboration, which limits our views to recognize their contributions to supply chain collaboration. In the current study, IOS appropriation has three distinctive components: IOS use for integration, IOS use for communication, and IOS use for intelligence.

Collaborative culture is considered as another important antecedent variable with four subcomponents: collectivism, long term orientation, power symmetry, and uncertainty avoidance. Collectivism and long term orientation are identified based on trust based rationalism. Power symmetry is viewed from resource dependence theory and social exchange theory. Uncertainty avoidance is evaluated based on transaction cost economics.

In explaining the important role of trust in supply chain collaboration, transaction cost economics argues that trust mitigates the probability of a firm's opportunistic behavior, which accounts for the risk in supply chain collaboration. As a complementary explanation, trust based rationalism also identifies trust as the indispensable antecedent to supply chain collaboration. In contrast to the negative assumption of transaction cost economics, trust based rationalism argues there are some supply chain partners who take the assumption of trustworthiness rather than opportunism in their collaboration with supply chain partners (Hill 1990; Hart and Saunders 1997). Trust based rationalism views trust rather than politics and conflicts as crucial to understanding interaction processes. Trust in itself is the key issue in IOS enabled supply chain collaboration.

Resource based view, relational view, extended resource based view, and social dilemma theories perceive collaborative advantage (i.e., joint competitive advantage) as the consequence of supply chain collaboration.

# References

Amit, R., & Schoemaker, P. (1993). Strategic assets and organizational rent. *Strategic Management Journal, 14*(1), 33–46.

Angeles, R., & Nath, R. (2001). Partner congruence in electronic data interchange (EDI) enabled relationships. *Journal of Business Logistics, 22*(2), 109–127.

Bagchi, P. K., & Skjoett-Larsen, T. (2005). Supply chain integration: A survey. *The International Journal of Logistics Management, 16*(2), 275–294.

Barney, J. (1991). Firm resources and sustained competitive advantage. *Journal of Management, 17*(1), 99–120.

Barringer, B. R., & Harrison, J. S. (2000). Walking a tightrope: Creating value through interorganizational relationships. *Journal of Management, 26*(3), 367–403.

Blau, P. M. (1964). *Exchange and power in social life.* New York: Wiley.

Bleeke, J., & Ernst, D. (1993). *Collaborating to compete: Using strategic alliances and acquisitions in the global marketplace.* New York: John Wiley and Sons.

Bromiley, P., & Cummings, L. L. (1995). Transaction costs in organizations with trust. In R. Bies, B. Sheppard, & R. Lewicki (Eds.), *Research on negotiation in organizations, 5* (pp. 219–250). Greenwich: JAI Press.

Cohen, W. D., & Levinthal, D. A. (1990). Absorptive capacity: A new perspective on learning and innovation. *Administrative Science Quarterly, 35*(1), 128–152.

Croom, S. (2001). Restructuring supply chains through information channel innovation. *International Journal of Operations & Production Management, 21*(4), 504–515.

Das, T. K., & Teng, B. S. (2000). A resource-based theory of strategic alliances. *Journal of Management, 26*(1), 31–62.

Das, T. K., & Teng, B. S. (2002). Social exchange theory of strategic alliances. In F. J. Contractor & P. Lorange (Eds.), *Cooperative Strategies and Alliances* (pp. 439–460). Oxford: Elsevier Science.

Dawes, R. M. (1980). Social dilemmas. *Annual Review of Psychology, 31*, 169–193.

Duffy, R., & Fearne, A. (2004). The impact of supply chain partnerships on supplier performance. *International Journal of Logistics Management, 15*(1), 57–71.

Dyer, J. H., & Singh, H. (1998). The relational view: Cooperative strategy and sources of interorganizational competitive advantage. *Academy of Management Review, 23*(4), 660–679.

Faulkner, D. O. (1995). *International strategic alliances: Co-operating to compete.* Maidenhead: McGraw-Hill.

Fawcett, S. E., Fawcett, A., Watson, B., & Magnan, G. (2012). Peeking inside the black box: toward an understanding of supply chain collaboration dynamics. *Journal of Supply Chain Management, 48*(1), 44–72.

Grabner, J., & Rosenberg, L. J. (1969). Communication in distribution channel systems. In L. W. Stern (Ed.), *Distribution channels: Behavior dimensions.* New York: Houghton-Mifflin Company.

Grover, V., Teng, J., & Fiedler, K. (2002). Investigating the role of information technology in building buyer-supplier relationships. *Journal of Association for Information Systems, 3*, 217–245.

Hamel, G. (1991). Competition for competence and inter-partner learning within international strategic alliances. *Strategic Management Journal, 12*(4), 83–103.

Hansen, M.H., Hoskisson, R., Lorenzoni, G., & Ring, P.S. (1997). Strategic capabilities of the transactionally-intense firm: Leveraging inter-firm relationships. *Working paper.* Texas A&M University

Harland, C. M., Zheng, J., Johnsen, T. E., & Lamming, R. C. (2004). A conceptual model for researching the creation and operation of supply networks. *British Journal of Management, 15*(1), 1–21.

Harrigan, K. R. (1988). Joint venture and competitive strategy. *Strategic Management Journal, 9*(2), 141–158.

Hart, P., & Saunders, C. (1997). Power and trust: Critical factors in the adoption and use of electronic data interchange. *Organization Science, 8*(1), 23–43.

Hill, C. W. (1990). Cooperation, opportunism, and the invisible hand: Implications for transaction cost theory. *Academy of Management Review, 15*, 500–513.

Hitt, M. (2011). Relevance of strategic management theory and research for supply chain management. *Journal of Supply Chain Management, 47*(1), 9–13.

Holweg, M., Disney, S., Holmström, J., & Småros, J. (2005). Supply chain collaboration: Making sense of the strategy continuum. *European Management Journal, 23*(2), 170–181.

Jap, S. D. (2001). Perspectives on joint competitive advantages in buyer-supplier relationships. *International Journal of Research in Marketing, 18*(1/2), 19–35.

Johnson, J. J., & Sohi, R. S. (2003). The development of interfirm partnering competence: Platforms for learning, learning activities and consequences of learning. *Journal of Business Research, 56*(9), 757–766.

Kaufman, A., Wood, C. H., & Theyel, G. (2000). Collaboration and technology linkages: A strategic supplier typology. *Strategic Management Journal, 21*(6), 649–663.

Khanna, T., Gulati, R., & Nohria, N. (1998). The dynamics of learning alliances: Competition, cooperation, and relative scope. *Strategic Management Journal, 19*, 193–210.

Kim, K. K., Umanath, N. S., & Kim, B. H. (2005). An assessment of electronic information transfer in B2B supply-channel relationships. *Journal of Management Information Systems, 22*(3), 293–320.

Kling, R. (1980). Social analysis of computing: Theoretical perspectives in recent empirical research. *ACM Computing Surveys, 12*(1), 61–110.

Knudsen, D. (2003). Aligning corporate strategy, procurement strategy and e-procurement tools. *International Journal of Physical Distribution and Logistics Management, 38*(8), 720–734.

Kogut, B. (1988). Joint ventures: Theoretical and empirical perspectives. *Strategic Management Journal, 9*(4), 319–332.

Koh, J., & Venkatraman, N. (1991). Joint venture formations and stock market reactions: An assessment in the information technology sector. *Academy of Management Journal, 34*(4), 869–892.

Kollock, P. (1998). Social dilemmas: Anatomy of cooperation. *Annual Review of Sociology, 24*, 183–214.

Kumar, R., & Nti, K. O. (1998). Differential learning and interaction in alliance dynamics: A process and outcome discrepancy model. *Organization Science, 9*(3), 356–367.

Kumar, K., & Van Dissel, H. G. (1996). Sustainable collaboration: Managing conflict and cooperation in interorganizational system. *MIS Quarterly, 20*(3), 279–300.

Kumar, K., Van Dissel, H. G., & Bielli, P. (1998). The merchant of Prato revisited: Toward a third rationality of information systems. *MIS Quarterly, 22*(2), 199–226.

Lambert, D.M., & Christopher, M.G. (2000). From the Editors. *International Journal of Logistics Management, 11*(2), pii–ii.

Lane, P. J., & Lubatkin, M. (1998). Relative absorptive capacity and interorganizational learning. *Strategic Management Journal, 19*(5), 461–477.

Lavie, D. (2006). The competitive advantage of interconnected firms: An extension of the resource-based view. *Academy of Management Review, 31*(3), 638–658.

Lee, H., Padmanabhan, V., & Whang, S. (1997). The bullwhip effect in supply chain. *Sloan Management Review, 38*(3), 93–102.

Lorenzoni, G., & Lipparini, A. (1999). The leveraging of interfirm relationships as a distinctive organizational capability: A longitudinal study. *Strategic Management Journal, 20*(4), 317–338.

Lundvall, B. A. (1992). *National systems of innovation: Towards a theory of innovation and interactive learning.* London: Pinter.

Macneil, I. R. (1980). *The new social contract: An inquiry into modern contractual relations.* New Haven: Yale University Press.

Malhotra, A., Gasain, S., & El Sawy, O. A. (2005). Absorptive capacity configurations in supply chains: Gearing for partner-enabled market knowledge creation. *MIS Quarterly, 29*(1), 145–187.

Manthou, V., Vlachopoulou, M., & Folinas, D. (2004). Virtual e-Chain (VeC) model for supply chain collaboration. *International Journal of Production Economics, 87*(3), 241–250.

March, J. G. (1995). The future, disposable organizations and the rigidities of imagination. *Organization, 2*, 427–440.

Min, S., Roath, A., Daugherty, P. J., Genchev, S. E., Chen, H., & Arndt, A. D. (2005). Supply chain collaboration: What's happening? *International Journal of Logistics Management, 16*(2), 237–256.

Mohr, J., Fisher, R. J., & Nevin, J. R. (1996). Collaborative communication in interfirm relationships: Moderating effects of integration and control. *Journal of Marketing, 60*(3), 103–115.

Mohr, J., & Nevin, J. R. (1990). Communication strategies in marketing channels: A theoretical perspective. *Journal of Marketing, 54*(4), 36–51.

Mohr, J., & Spekman, R. E. (1994). Characteristics of partnership success: partnership attributes, communication behavior, and conflict resolution techniques. *Strategic Management Journal, 15*(2), 135–152.

Mowery, D.C., Oxley, J.E., & Silverman, B.S. (1996). Strategic alliances and interfirm knowledge transfer. *Strategic Management Journal, 17*(Winter Special Issue), 77–91.

Nesheim, T. (2001). Externalization of the core: Antecedents of collaborative relationships with suppliers. *European Journal of Purchasing and Supply Management, 7*(4), 217–225.

Park, N. K., Mezias, J. M., & Song, J. (2004). A resource-based view of strategic alliances and firm value in the electronic marketplace. *Journal of Management, 30*(1), 7–27.

Pfeffer, J., & Nowak, P. (1996). Joint ventures and interorganizational interdependence. In P. J. Buckley & J. Michie (Eds.), *Firms, organizations and contracts: A reader in industrial organization* (pp. 385–409). New York: Oxford University Press.

Prahalad, C. K., & Hamel, G. (1990). The core competence of the corporation. *Harvard Business Review, 68*(3), 79–91.

Ragatz, G. L., Handfield, R. B., & Scannell, T. V. (1997). Success factors for integrating suppliers into new product development. *Journal of Product Innovation Management, 14*(3), 190–202.

Sabath, R. E., & Fontanella, J. (2002). The unfulfilled promise of supply chain collaboration. *Supply Chain Management Review, 6*(4), 24–29.

Scott, J. (1987). *Organizations.* Englewoods Cliffs: Simon and Schuster.

Sheu, C., Yen, H.R., & Chae, D. (2006). Determinants of supplier-retailer collaboration: Evidence from an international study. *International Journal of Operations and Production Management, 26*(1), 24–49

Simatupang, T. M., & Sridharan, R. (2002). The collaborative supply chain. *International Journal of Logistics Management, 13*(1), 15–30.

Simatupang, T. M., & Sridharan, R. (2005). The collaboration index: A measure for supply chain collaboration. *International Journal of Physical Distribution and Logistics Management, 35*(1), 44–62.

Simonin, B. L. (1997). The importance of collaborative know-how: An empirical test of the learning organization. *Academy of Management Journal, 40*(2), 1150–1174.

Slater, S. F., & Narver, J. C. (1995). Market orientation and the learning organization. *Journal of Marketing, 59*(3), 63–74.

Son, J., Narasimhan, S., & Riggins, F. J. (2005). Effects of relational factors and channel climate on EDI Usage in the customer-supplier relationship. *Journal of Management information Systems, 22*(1), 321–353.

Spekman, R. E., Forbes, T. M., Isabella, L. A., & MacAvoy, T. C. (1998). Alliance management: A view from the past and look to the future. *Journal of Management Studies, 35*(6), 747–772.

Stank, T. P., Daugherty, P. J., & Ellinger, A. E. (1999). Marketing/Logistics integration and firm performance. *International Journal of Logistics Management, 10*(1), 11–23.

Stank, T. P., Keller, S. B., & Daugherty, P. J. (2001). Supply chain collaboration and logistical service performance. *Journal of Business Logistics, 22*(1), 29–48.

Subramani, M. (2004). How do suppliers benefit from information technology use in supply chain relationships? *MIS Quarterly, 28*(1), 45–73.

Teece, D. J., Pisano, P. G., & Shuen, A. (1997). Dynamic capabilities and strategic management. *Strategic Management Journal, 18*(7), 509–533.

Teo, H. H., Wei, K. K., & Benbasat, I. (2003). Predicting intention to adopt interorganizational linkages: An institutional perspective. *MIS Quarterly, 27*(1), 19–49.

Thomas, D., & Ranganathan, C. (2005). Enabling e-business transformation through alliances: Integrating social exchange and institutional perspectives. In *Proceedings of the 38th Hawaii International Conference on System Sciences.*

Thorelli, H. B. (1986). Networks: Between markets and hierarchies. *Strategic Management Journal, 7*(1), 37–51.

Uzzi, B. (1997). Social structure and competition in interfirm networks: The paradox of embeddedness. *Administrative Science Quarterly, 42*(1), 35–67.

Verdecho, M., Alfaro-Saiz, J., Rodriguez–Rodriguez, R., & Ortiz-Bas, A. (2012). A multi-criteria approach for managing inter-enterprise collaborative relationships. *Omega, 40*(3), 249–263.

Verwaal, E., & Hesselmans, M. (2004). Drivers of supply network governance: An explorative study of the dutch chemical industry. *European Management Journal, 22*(4), 442–451.

Volberda, H. W. (1998). *Building the flexible firm: How to remain competitive.* Oxford: Oxford University Press.

Von Hippel, E. (1988). *The sources of innovation.* New York: Oxford University Press.

Wever, M., Wongnum, P., Trienekens, J., & Omta, S. (2012). Supply chain-wide consequences of transaction risks and their contractual solutions: Towards an extended transaction cost economics framework. *Journal of Supply Chain Management, 48*(1), 73–91.

Williamson, O. E. (1975). *Markets and hierarchies.* Englewood Cliffs: Prentice-Hall.

Wuyts, S., & Geyskens, I. (2005). The formation of buyer–supplier relationships: Detailed contract drafting and close partner selection. *Journal of Marketing, 69*(4), 103–117.

Zeng, M., & Chen, X. P. (2003). Achieving cooperation in multiparty alliances: A social dilemma approach to partnership management. *Academy of Management Review, 28*(4), 587–605.

# Chapter 3
# Antecedents of Supply Chain Collaboration

**Abstract** In researching the antecedents or the conditions that lead to or affect supply chain collaboration, prior studies focus on IT capability and the use of interorganizational systems (IOS) but simplify or ignore its context (e.g., culture and trust). Although IT capability and IOS use are necessary for supply chain collaboration to succeed, organizational culture and trust must be taken into consideration simultaneously. Many supply chain collaborations fail due to incompatible corporate culture, distrust, and the complexities involved. In this chapter, we identify, define, and elaborate four antecedents of supply chain collaboration: IT resources, IOS appropriation, collaborative culture, and trust. Specifically, IT resources include IT infrastructure flexibility and IT expertise. IOS appropriation has been differentiated as IOS use for integration, IOS use for communication, and IOS use for intelligence. Four elements of collaborative culture (i.e., collectivism, long-term orientation, power symmetry, and uncertainty avoidance) and two dimensions of trust (i.e., credibility and benevolence) are investigated. Before developing and testing the relationships in the proposed framework, it is theoretically and conceptually sound to carefully identify, define, and discuss the key antecedent constructs in the framework through a review of literature and discussion of theoretical logic in the following sections.

## 3.1 IT Resources

In information systems literature, IT resources are defined as a firm's ability to deploy IT based resources "in combination or copresent with other resources and capabilities" (Bharadwaj 2000, p. 171) and "to affect a predetermined outcome" (McKeen et al. 2005, p. 662). King (2002) views IT resources as bundles of internally consistent elements that are focused toward the fulfillment of an IT or business objective. Piccoli and Ives (2005) and Wade and Hulland (2004) argue

M. Cao and Q. Zhang, *Supply Chain Collaboration*,
DOI: 10.1007/978-1-4471-4591-2_3, © Springer-Verlag London 2013

that IT resources encompass IT assets (i.e., anything a firm can use in offering its products) and IT capabilities (i.e., ability to mobilize IT assets).

Most researchers use resource based view to explain IT resources, IT assets, and IT capabilities (Bharadwaj 2000; Bhatt and Grover 2005; Ravichandran and Lertwongsatien 2005; Teo et al. 2011; Rai et al. 2012; Grover and Kohli 2012) but they do not strictly distinguish between these concepts. Even if some researchers have tried to distinguish these concepts conceptually, they mix them when conceptualizing or operationalizing their subcomponents. Due to the intangible and abstract nature of these concepts, they are difficult to operationalize. In previous studies, IT resources are studied within the context of individual firms (Rockart et al. 1996; Ross et al. 1996; Bharadwaj 2000; Santhanam and Hartono 2003; Ravichandran and Lertwongsatien 2005). To the researcher's knowledge, few studies have been conducted to conceptualize IT resources in the context of IOS enabled supply chain collaboration. In current research, IT resources are defined as the bundles of IT assets and capabilities that can be used to support IOS use in supply chain collaboration.

Researchers agree that IT resources are a multidimensional concept with two common components: IT infrastructure flexibility and IT expertise (i.e., technical IT skills and management IT knowledge). Ravichandran and Lertwongsatien (2005) identify three broad categories of IT resources in the IS literature: human, technological, and relationship resources and propose a research model incorporating IS human capital, IT infrastructure flexibility, and IS relationship quality. Ross et al. (1996) also classify three types of IT assets that constitute a firm's IT resources: human, technology, and relationship.

Bharadwaj (2000) maintains that IT resources include IT infrastructure, IT human resources (i.e., technical and managerial IT skills), and intangible IT based resources (i.e., knowledge assets, customer orientation, and synergy). Melville et al. (2004) categorize technological IT resources into IT infrastructure (i.e., shared technology and technology services across the organization) and specific business applications that utilize the infrastructure (e.g., purchasing systems and CFPR tools) (Broadbent and Weill 1997). Another IT resource that they identify is the firm's human capital including expertise and knowledge.

Peppard and Ward (2004) portray IS capability as having three inter-related attributes: a fusion of business knowledge with IS knowledge, a flexible and reusable IT platform, and an effective use process. Bhatt and Grover (2005) classify IT capability into two categories: value capability (i.e., IT infrastructure) and competitive IT capability (i.e., IT management capabilities). IT management capability further includes IT business experience (the extent to which IT groups understand business) and relationship infrastructure (the extent to which there are positive relationships between IT and business managers).

The literature review demonstrates many IT enabled intangibles can be included in the conceptualization of IT resources. In current research, IT resources consist of two most common components: IT infrastructure flexibility and IT expertise (Table 3.1).

**Table 3.1** Definition of IT resources and subcomponents

| Construct | Definition | Literature |
|---|---|---|
| IT resources | The bundles of IT assets and capabilities that can be used to support IOS use in supply chain collaboration | Bharadwaj (2000), Bhatt and Grover (2005), King (2002), Melville et al. (2004), McKeen et al. (2005), Peppard and Ward (2004), Piccoli and Ives (2005), Ravichandran and Lertwongsatien (2005), Ross et al. (1996) |
| IT infrastructure flexibility | The extent to which systems (i.e. hardware, software, communication technologies, and database) are easily reconfigurable to support different business applications and services | Armstrong and Sambamurthy (1999), Bharadwaj (2000), Broadbent and Weill (1997), Byrd and Turner (2000), Davenport and Linder (1994), Duncan (1995), Piccoli and Ives (2005), Ray et al. (2005), Ross et al. (1996), Weill et al. (1996) |
| IT expertise | The extent to which IT staff and managers are able to provide technical and business solutions | Bharadwaj (2000), Dehning and Richardson (2002), Melville et al. (2004), McKenney et al. (1995), Piccoli and Ives (2005), Ranganathan et al. (2004), Ravichandran and Lertwongsatien (2005), Ross et al. (1996) |

## *3.1.1 IT Infrastructure Flexibility*

IT infrastructure flexibility refers to the extent to which systems (i.e. hardware, software, communication technologies, and database) are easily reconfigurable to support different business applications and services. IT infrastructure comprises information and communication technologies as well as shared technical platforms and databases (Ross et al. 1996; Weill et al. 1996; Bharadwaj 2000). The primary constituents of IT infrastructure are computing platform (hardware and operating systems), communications network, critical shared database, and core applications (Byrd and Turner 2000). IT infrastructure is the foundation of IT assets (i.e., technical and human assets) and services shared across a firm (Piccoli and Ives 2005). As a result, IT infrastructure provides shared foundation for the delivery of business applications and services (Broadbent and Weill 1997).

IT infrastructure has been identified as the capabilities that influence a firm's ability to use IT strategically (Weill 1993; Davenport and Linder 1994; Duncan 1995; Ross et al. 1996; Armstrong and Sambamurthy 1999; Broadbent et al. 1999; Sambamurthy et al. 2003; Ray et al. 2005). Peppard and Ward (2004) claim that a flexible and reusable IT platform not only provides the technical platform, services, and resources needed to quickly respond to business changes but also provides the capacity to develop innovative applications supporting new processes or business initiatives. IT infrastructure varies in reach (the extent of the connectivity) and range (the scope of services) (Keen 1991). As reach and range increase, the resources made available by IT infrastructure and the ability to support a variety of strategic initiatives will increase as well (Broadbent et al. 1999). In this sense, flexible IT infrastructure is a valuable capability to support IOS use in supply chain collaboration. The flexibility nature of IT infrastructure is manifested in the extent to which a firm adopts systems with standards, modularity, compatibility, and scalability. Systems with such characteristics make it easier for data and applications to be shared and accessed throughout the organization and across the firm boundaries (Broadbent and Weill 1997; Ray et al. 2005).

## 3.1.2 IT Expertise

IT expertise refers to the extent to which IT staff and managers are able to provide technical and business solutions. IT expertise is an important input in the development of IT resources (Ravichandran and Lertwongsatien 2005). It denotes technical IT skills (e.g., application development, systems integration, and systems maintenance) and managerial IT knowledge (e.g., ability to work with other business units and external organizations, recognize and select projects, gather and allocate resources, and lead development teams) (Ross et al. 1996; Bharadwaj 2000; Dehning and Richardson 2002; Melville et al. 2004).

Technical IT skills refer to the general skills, experience, and expertise (e.g., programming, network, Web development) possessed by IT staff to design and develop effective applications and systems. As such, technical IT skills include proficiency in system analysis and design, programming, infrastructure design, etc. (McKenney et al. 1995; Ross et al. 1996; Piccoli and Ives 2005). Although it is argued that technical IT skills are easily obtainable on the market (Mata et al. 1995; Ray et al. 2005), they are subject to organizational learning dynamics and knowledge barriers because IT activities are generally considered knowledge intensive and require specific technical skills (Attewell 1992; Fichman 2000; Piccoli and Ives 2005). Thus, existing particular knowledge or wide-ranging technical skill sets allow firms to adopt and use IT more easily (Cohen and Levinthal 1990). Firms that have highly skilled IT personnel are better positioned to develop higher level of IT resources than those that do not (Ravichandran and Lertwongsatien 2005).

Managerial IT knowledge refers to the combination of IT-related and business-related knowledge possessed and exchanged by IT staff and managers (Ranganathan et al. 2004). Specifically, it includes the ability to lead the IS function, manage IT projects, evaluate technology options, manage change, and envision creative and feasible technical solutions to business problems (Mata et al. 1995; McKenney et al. 1995; Ross et al. 1996; Feeny and Willcocks 1998; Piccoli and Ives 2005). Rockart (1988) believes that shared knowledge between managers determines the strategic use of IT. Boynton et al. (1994) propose that IT use in an organization is influenced by the mixture of IT-related knowledge of managers. Managerial IT knowledge and skills can significantly reduce the costs and lead time associated with IT development (Bharadwaj 2000). IT skills are developed through the process of organizational learning (Piccoli and Ives 2005). Drawing on the resource based view, Mata et al. (1995) recognize managerial IT skills as a source of sustained competitive advantage.

## 3.2   IOS Appropriation

Interorganizational systems (IOS) or information technology applications that span firm boundaries have been extensively studied in IS literature (Massetti and Zmud 1996; Subramani 2004; Saeed et al. 2005). IOS refer to the information technology applications used to mediate buyer–supplier transactions and relationships (Subramani 2004). Barret and Konsynski (1982) use the term "interorganizational information sharing systems" for the first time. Cash and Konsynski (1985) define IOS as automated information systems shared by two or more companies. In a broad sense, IOS consist of computer and communications infrastructure for managing interdependencies between firms (Chi and Holsapple 2005). Premkumar (2000) views IOS as application systems that link various partners in the supply chain using a public or private telecommunication infrastructure to provide computer-to-computer communication of business transactions and documents. IOS are now used to enable cooperation more than competition among firms (Hong 2002). They are perceived as cooperative endeavors between otherwise independent organizations (Kumar and van Dissel 1996).

IOS literature reveals multiple goals motivating their use: necessity (meeting regulatory requirements), asymmetry (exerting power or control over other firms), reciprocity (pursuing mutual benefits), efficiency, agility, innovation, stability, and legitimacy (Oliver 1990; Premkumar et al. 1997; El Sawy et al. 1999; Chi and Holsapple 2005). To explain diverse outcomes, IOS use has been conceptualized as breadth, depth, intensity, volume, scope, and diversity (Bensaou and Venkatraman 1995; Massetti and Zmud 1996; Saeed et al. 2005; Teo et al. 2011). However, these definitions fail to express clearly the purpose or intentionality of IOS use and thus are not that useful in capturing the use of IOS motivated by different goals (Subramani 2004). Subramani (2004) labels the patterns of IT use as IT appropriation, which is consistent with the notion of DeSanctis and Poole (1994), Chin et al. (1997) and Salisbury et al. (2002). Subramani (2004) claims different IT appropriations can bring about different outcomes although the underlying technologies and the context of IT use are similar. The current research adopts their views and defines IOS appropriation as patterns, modes, or fashions of IOS use.

In examining the impact of IT on interfirm relations and the modes of governance, Malone et al. (1987) classify the impact of IT into electronic communication effects (i.e., reducing cost of communication while expanding reach) and electronic integration effects (i.e., increasing the degree of interdependence between partners by creating joint, interpenetrating processes). Saeed (2004) develops a research model that posits IOS characteristics as the main antecedent to supply chain integration whereby IOS characteristics include IOS integration and IOS intelligence. By synthesizing their work and adapting them into the context of supply chain collaboration, the current research introduces three components of IOS appropriation: IOS use for integration, IOS use for communication, and IOS use for intelligence (Table 3.2). These three components support real time information sharing in supply chain collaboration. Furthermore, they have their own focuses and play different roles in collaboration between supply chain partners: enhancing process integration, facilitating communication, and enabling learning and knowledge creation.

## 3.2.1 IOS Use for Integration

IOS use for integration refers to the extent of IOS use in facilitating electronic process coupling between supply chain partners. The tight coupling of processes enables idiosyncratic and mutually dependent partners to form the unified whole (Barki and Pinsonneault 2005). IOS technologies and applications for integration involve managing customer–supplier relationships, e.g. EDI, CPFR, ECR, VMI, Web-based procurement systems, electronic trading systems, RFID, customer relationship management (CRM), supply chain management systems, enterprise resource planning (ERP), Internet/Intranet/Extranet, portals, e-hubs, workflow automation (e.g., Computer Aided Design (CAD), Computer Aided Manufacturing (CAM)), collaborative authoring, computer conferencing, and standards such as Rosettanet.net and Covisint.net. These IOS technologies and applications provide different levels of integration: information sharing (e.g., order, inventory) and collaborative planning (Kulp et al. 2004).

Electronic integration is an important impact of using IOS (Saeed et al. 2005). It means that trading partners use IT to create joint, interpenetrating processes (Malone et al. 1987; Kekre and Mukhopadhyay 1992; Hart and Saunders 1997; Grover et al. 2002). It is a strategic choice made by firms to transform business scopes or business networks by using information technologies to reengineer key business processes and business relations (Kambil and Short 1994). Electronic linkages are described as different ways that firms manage economic interdependence across value adding roles in the network of supply chain partners. Barua et al. (2004) define electronic/systems integration as the extent to which a firm integrates its IT systems to provide information visibility to partners to support online transactions across the supply chain. Bensaou and Venkatraman (1995) propose electronic interdependence as an interorganizational configuration that entails extensive use of IT in facilitating information sharing and collaborative processes in dyadic linkages. Christiaanse and Venkatraman (2002) conclude that a firm can enhance electronic integration by leveraging certain characteristics of IOS that enable it to monitor and direct the behavior of firms in the distribution channel. Manthou et al. (2004) contend that successful operations of supply chain

**Table 3.2** Definition of IOS appropriation and subcomponents

| Construct | Definition | Literature |
|---|---|---|
| IOS appropriation | Patterns, modes, or fashions of IOS use | Chin et al. (1997), DeSanctis and Poole (1994), Malone et al. (1987), Saeed (2004), Salisbury et al. (2002), Subramani (2004) |
| IOS use for integration | The extent of IOS use in facilitating electronic process coupling between supply chain partners | Barki and Pinsonneault (2005), Barua et al. (2004), Bensaou and Venkatraman (1995), Christiaanse and Venkatraman (2002), Grover et al. (2002), Hart and Saunders (1997), Kulp et al. (2004), Manthou et al. (2004), Mukhopadhyay and Kekre (2002), Saeed et al. (2005), Thomas and Ranganathan (2005) |
| IOS use for communication | The extent of IOS use in facilitating contacts and message flows between supply chain partners | Bafoutsou and Mentzas (2002), Chi and Holsapple (2005), Hill and Scudder (2002), Malone et al. (1987) |
| IOS use for intelligence | The extent of IOS use in enhancing learning and knowledge creation between supply chain partners | Chi and Holsapple (2005), Collins et al. (1998), Gini and Boddy (1998), Mehra and Nissen (1998), Milton et al. (1999), Nissen and Sengupta (2006), O'Leary (2003), Tsui (2003), Wurman et al. (1998) |

partnerships mandate that every member must be able to share information with trading partners in real-time, which is realized by enabling disparate information systems to share data in the context of specific business processes.

IOS use for integration falls within the realm of idiosyncratic interfirm linkages that entail close collaboration among business partners (Frohlich and Westbrook 2001). IOS use can tighten the coupling of processes that creates and uses information (Malone et al. 1987). For example, CAD/CAM technology allows design and manufacturing engineers in both supplier's and buyer's companies to access their respective data to test alternative designs and to create better products. Systems linking the supplier's and buyer's inventory management processes enable just-in-time delivery, and thus reduce the total inventory costs for the linked partners. Studies show that when EDI is used to closely couple operations between firms, it helps promote long-term collaboration because of relationship-specific assets/investments and high switching costs (Mukhopadhyay et al. 1995; Mukhopadhyay and Kekre 2002). Although companies could use a variety of supply chain connectivity mechanisms, EDI will continue to be used in combination with newer Internet-based technologies (Angeles and Nath 2001). Furthermore, the cost effectiveness of the newer Internet-based version of EDI will encourage more firms (large and small alike) to deploy newer technologies and thus participate in e-business (Droge and Germain 2000).

A body of literature is emerging on electronic integration enabled by Internet technologies and Web-based information systems. Historically there has been no ubiquitous, common network platform over which to share information until the emergence of the Internet (Manthou et al. 2004). Web technologies and the Internet enable supply chain partners to perform digital business operations better, faster, and cheaper than ever before. Various functionalities of Web-based systems support search, processing, monitoring and control, and coordination activities (Subramaniam and Shaw 2002). In fact, Web presence and e-business operations have become more of a competitive necessity for most supply chain members (Thomas and Ranganathan 2005). Zhu and Kraemer (2002) offer the concept of electronic commerce capabilities and argue that such capabilities are reflected in electronic commerce system functionalities and range from online order information, digital product catalogs, to integration with supplier databases. Mukhopadhyay and Kekre (2002) identify the strategic and operational benefits of electronic integration in B2B context. Chang and Shaw (2004) observe that a number of universal, extensible mark-up language (XML)-based process standards have been developed for supply chain collaboration initiatives, e.g., ebXML initiatives and the RosettaNet consortium. A variety of Internet-based coordination mechanisms have enhanced supply chain management through information sharing and process integration across the supply chain (Garcia-Dastugue and Lambert 2003; Lejeune and Yakova 2005). Overall, the literature review provides broad support for the important role of IOS in supporting various interorganizational activities, processes, and collaboration.

## 3.2.2 IOS Use for Communication

IOS use for communication refers to the extent of IOS use in facilitating contacts and message flows between supply chain partners. IOS technologies and applications for inter-firm communication include message services, channel management, communications network, and communication standards and protocols (Chi and Holsapple 2005). Examples of message services are email, fax, instant messaging, voice mail, electronic bulletin board, and controlled posting (e.g. FAQs). Call center, electronic funds transfer, point of sales (PoS), Web site, wireless device are technologies for channel management between supply chain partners. Communications network consists of peer-to-peer, broadband, intranet, extranet, Internet, and wireless networks. Communication standards and protocols comprise EDI, XML, Web services description language (WSDL), universal description, discovery, and integration (UDDI).

Web technologies and electronic networks have created an environment where communications between partners are extremely easy and fast. The use of message-based systems such as email, fax, instant messaging, and bulletin board enable frequent, bidirectional, and rich contact and communication between partners. Call centers, electronic funds transfer, PoS, Web sites, and wireless devices provide multiple communication channels, and some can directly transmit information to partner's applications resulting in fast and real-time contacts and message flows (McLaren et al. 2004). With the connectivity provided by advanced e-collaboration tools, e.g., electronic discussion groups, groupware, teamware, and electronic conferencing, supply chain partners can work together anytime, anywhere. e-Collaboration tools can bring geographically dispersed people together for virtual meetings across great distance, resulting in improved communication flows across organizations as well as faster and better decision making (Bafoutsou and Mentzas 2002). Hill and Scudder (2002) discuss that Web-based technologies can facilitate frequent and automatic bidirectional information flows between supply chain partners and thus enhancing the degree of collaboration between them. So, high level of IOS use for communication will greatly facilitate the collaboration between supply chain partners.

### 3.2.3 IOS Use for Intelligence

IOS use for intelligence refers to the extent of IOS use in enhancing learning and knowledge creation between supply chain partners. IOS technologies and applications for inter-firm intelligence could be shared data warehouse and data/text mining, shared repository database and decision support systems, shared digital documents and archives, shared knowledge acquisition, retrieval, and navigation, knowledge search (e.g. expert finder tool, meta/Web-crawler, taxonomy/ontological tools), knowledge discovery and generation analytics (e.g. online analytical processing (OLAP), simulation, modeling), artificial intelligence (e.g. intelligent agents, case-based reasoning, neural networks, genetic algorithm, and rule engines), group decision support systems, and software agents.

IOS use for intelligence gathering and analysis captures an organization's ability to facilitate joint learning and decision making, assimilate knowledge and skills from its partners, and jointly create new knowledge based on shared data repositories by using information technologies (Milton et al. 1999; Tsui 2003). It is similar to knowledge sharing receptivity (Chi and Holsapple 2005), assimilative ability (O'Leary 2003), or partner-specific absorptive capacity (Dyer and Singh 1998). It involves implementing a set of IOS or interorganizational processes that allow supply chain partners to systematically identify valuable know-how's and discover new knowledge, and then transfer them across organizational boundaries (Dyer and Singh 1998). Useful knowledge and intelligence may be buried in huge data repository and digital documents. By actively implementing knowledge systems, e.g., codifying, storing, structuring, filtering, integrating, retrieving, and transferring of usable knowledge assets, supply chain partners can integrate fragmented information, assimilate it, and thus jointly create value.

Many researchers emphasize the importance of using intelligence/knowledge agents and systems in supply chain collaboration (Collins et al. 1998; Gini and Boddy 1998; Mehra and Nissen 1998; Rodriguez-Aguilar et al. 1998; Wurman et al. 1998; Nissen and Sengupta 2006). Wurman et al. (1998) contend that intelligent software agents provide great potential for automation and support of supply chain processes. From the perspectives of the intermediation economics and agent technologies, Nissen (2000) analyzes the role of agent-based IT in supply chain disintermediation versus reintermediation. Caridi et al. (2005) find out that there are some hurdles that arose in implementing CPFR, signifying a strong need for providing collaboration process with an intelligent tool to optimize negotiation. Chung et al. (2005) hold that information overload often hinders knowledge discovery because the existing tools lack analysis and visualization capability. Nissen and Sengupta (2006) contend that software agents combine capabilities of several IT classes (e.g., DSS, expert systems, parallel processing, mobile computing) and are moving the boundaries of computer-aided decision making, e.g., autonomous, mobile decision makers. Thus IOS use for intelligence enables automation, knowledge discovery, and real-time decision making.

## 3.3  Collaborative Culture

Culture is not an individual's characteristic but an organizational trait (Hofstede 1998; Ashenbaum et al. 2012). Organizational culture is widely studied in the management literature and is often cited as a cause for the failure of interorganizational collaborative relationships (Segil 1998; Kumar et al. 1998; Gopal and Gosain 2010). Schein (1985) defines organizational culture as a set of basic assumptions developed by the organization as it learns to deal with problems within the organization and changes in its external environment. Gregory (1983) regards organizational culture as the shared meaning among people (e.g., role expectation, how to solve problems, and authority structure). It is the mental model of all members of the organization. It describes the multifaceted set of knowledge that organizational members use to perform tasks and generate social behaviors (Reichers and Schneider 1990; Bates et al. 1995).

In this research, organizational culture refers to the norms, beliefs, and underlying values shared in a firm regarding appropriate business practices in the supply chain (Nooteboom et al. 1997; Boddy et al. 2000; Wuyts and Geyskens 2005). Organizational culture may encourage or discourage collaboration in the context of partnering (Boddy et al. 2000). Collaborative culture deals with a relationship orientation where the primary emphasis is put on maintaining long-term relationships, even sometimes the organizational goals have to be modified to avoid harms to the partnership (Walls 1993; Kumar et al. 1998). Collaborative culture is defined as the norms, beliefs and underlying values with relationship orientation shared in a firm regarding appropriate business practices in the supply chain (Walls 1993; Kumar et al. 1998; Boddy et al. 2000; Wuyts and Geyskens 2005). Firms with

**Table 3.3** Definition of collaborative culture and subcomponents

| Construct | Definition | Literature |
|---|---|---|
| Collaborative culture | The norms, beliefs and underlying values with relationship orientation shared in a firm regarding appropriate business practices in the supply chain | Bates et al. (1995), Boddy et al. (2000), Gregory (1983), Hofstede (1998), Kumar et al. (1998), Nooteboom et al. (1997), Reichers and Schneider (1990), Schein (1985), Segil (1998), Walls (1993), Wuyts and Geyskens (2005) |
| Collectivism | The extent to which a firm holds "we" rather than "I" consciousness when working with supply chain partners | Hofstede (2000), Min et al. (2005), Sako and Helper (1998), Steensma et al. (2000), Wuyts and Geyskens (2005) |
| Long term orientation | The extent to which a firm is willing to exert efforts in developing an enduring relationship with supply chain partners | Angeles and Nath (2001), Axelrod (1984), Cachon and Lariviere (2001), Dyer (1996), Hofstede (2001), Holweg et al. (2005), Schultze and Orlikowski (2004), Sheu et al. (2006) |
| Power symmetry | The extent to which a firm believes that supply chain partners should have an equal say in their relationships | Bates et al. (1995), Gundlach and Hofstede (1980), McAlister et al. (1986), Narayandas and Rangan (2004), Porter (1980), Son et al. (2005), Tuten and Urban (2001), Verwaal and Hesselmans (2004), Wuyts and Geyskens (2005) |
| Uncertainty avoidance | The extent to which a firm feels threatened by and tries to evade ambiguous situations in the supply chain | Bensaou and Venkatraman (1995), Coase (1988), Dyer and Singh (1998), Fransman (1994), Wuyts and Geyskens (2005), Hofstede (2001), Kaufman et al. (2000), Kim et al. (2005), Steensma et al. (2000) |

collaborative culture are more likely to coordinate with their supply chain partners based on trust, good will, and social norms rather than impersonal and legal contracts, firm rules, and fixed goals.

To have a more comprehensive view of supply chain collaboration, organizational culture, as an important organizational context, must be incorporated into the understanding of the phenomenon (Orlikowski 1993). Four elements of collaborative organizational culture are investigated: collectivism, long-term orientation, power symmetry, and uncertainty avoidance (Table 3.3). They are firm-level equivalents of the national-level dimensions proposed by Hofstede (1980, 1991). Hofstede's (1980) another dimension, masculinity, is not included in this study because it is difficult to adapt it to the supply chain context. Kumar et al. (1998) have tried to tailor masculinity to the firm level as earning power and dominance, which is captured by the dimension of power symmetry in this study.

### 3.3.1  Collectivism

Collectivism refers to the extent to which a firm holds "we" rather than "I" consciousness when working with supply chain partners (Hofstede 1980, 1991). Collectivists value social fabric and norms rather than individual objectives (Steensma et al. 2000), and thus collectivists are more cooperative. They emphasize group and collective contributions to the collaboration (Bates et al. 1995). Collectivists enjoy working together and coordinating each other's efforts. They care about their business partners and thus perform better in close cooperation with partners (Hofstede 2001; Wuyts and Geyskens 2005).

Firms with collectivism orientation are more likely to form cooperative partnerships, encourage teamwork, exchange information between partners, and solve problems jointly (Wagner 1995). Individualist firms value the independence and flexibility provided by the arm's length relationship and prefer formal contracts as a mechanism for conflict resolution (Steensma et al. 2000; Wuyts and Geyskens 2005). In contrast, collectivists focus more on cooperation and joint efforts with a feeling of "we are in this together" (Min et al. 2005). When collectivists make decisions, both parties are taken into account. They pursue relational governance and prefer imprecise contracts that commit parties together to solve difficulties as they emerge (Sako and Helper 1998). Collectivism creates a sense of duty in relationships and a dislike of arm's length bargaining relationship (Steensma et al. 2000).

### 3.3.2  Long Term Orientation

Long term orientation refers to the extent to which a firm is willing to exert efforts in developing an enduring relationship with supply chain partners (Sheu et al. 2006). Long-term orientation or commitment is often cited as the predictor for successful interorganizational relationships (Angeles and Nath 2001; Schultze and Orlikowski 2004; Holweg et al. 2005). It is demonstrated by committing resources to the relationships (e.g., technologies, time, money, and facilities) (Sheu et al. 2006). The willingness of supply chain partners to maintain long-term relationships is also demonstrated by being of assistance during difficult times or when diverging interests arise (Angeles and Nath 2001). Supply chain partners should overcome diverse short-term interests and unselfishly work together because conflicts of interests mitigate the commitment of relationship-specific investment, information sharing, and supply chain collaboration (Cachon and Lariviere 2001; Holweg et al. 2005).

Long-term orientation depends on the firm's willingness to establish a long-term relationship and make relationship-specific investment (Sheu et al. 2006). Successful supply chain partnerships will be nurtured when parties involved show willingness to work together for long time and commit diverse assets to future transactions (Dyer 1996). It is the expectation of enormous and endless future interactions that encourage partners to cooperate for their mutual gains (Schultze and Orlikowski 2004). Therefore, the relationship is governed not by a formal contract but by an implicit social contract because partners cooperate out of mutual obligations (Axelrod 1984; Schultze and Orlikowski 2004). When firms make transaction- or relation- specific investments, process efficiency and productivity will be improved and the collaboration between partners will be enhanced (Dyer 1996; Bensaou and Anderson 1999).

### 3.3.3 Power Symmetry

Power symmetry refers to the extent to which a firm believes that supply chain partners should have an equal say in their relationships. Power symmetry means low power distance. Power distance is the practice of inequalities in distributing power and authority among partners (Hofstede 1980). Firms with low power distance are more likely to participate in equality and consultative decision making, while those with high power distance are more likely to operate based on authority and explicit definition of tasks (Hofstede 1980; Bates et al. 1995; Wuyts and Geyskens 2005; Vegt et al. 2010). Supply chain partners are normally not equal in terms of clout and bargaining power (Min et al. 2005). Min et al. (2005) suggest the powerful firm not take advantage of its position but try to meet the less powerful partner's needs in mutually beneficial arrangements, even though its partner is captive.

A strong relationship is often related to an equal balance of power (Tuten and Urban 2001). Firms with low power distance view their supply chain partners as relatively equal and engage in informal communication with partners at different levels (Hofstede 1980). The governance is often based on shared values, or a sense of duty, or obligation to others (Wuyts and Geyskens 2005). If one tries to over-power another partner, it will cause conflicts between them and thus partnering will fail. Son et al. (2005) observe that exercising bargaining power through coercive influence may decrease positive attitudes toward the firm and thus it has an unfavorable effect on building cooperative and long-lasting interfirm relationships.

Scholars have concluded that asymmetrical power and dependence result in dysfunctional relationships (McAlister et al. 1986; Gundlach and Ernest 1994; Verwaal and Hesselmans 2004). The imbalance of power engenders asymmetrical relationship whereby powerful parties dictate to weaker parties and extract returns in proportion to their influence (Porter, 1980; McAlister et al. 1986; Narayandas and Rangan 2004). As such, the collaborative relationship will not sustain long. Partners' switching costs are going down with Web-based advanced EDI technologies. If the powerful firm does not treat its weak partners as equal, its partners will go away and switch to other collaborators. Long-term relationships have to be motivated by the mutuality of intent and benefit sharing (Angeles and Nath 2001). Power symmetry plays a greater role in supporting more democratic and participative partner relationships. Narayandas and Rangan (2004) contend that power asymmetry can be redressed through the development of trust and inter-organizational commitment.

## 3.3.4 Uncertainty Avoidance

Uncertainty avoidance refers to the extent to which a firm feels threatened by and tries to evade ambiguous situations in the supply chain (Hofstede 2001; Wuyts and Geyskens 2005). Firms vary in their tolerance of uncertainty and ambiguity (Wuyts and Geyskens 2005). Firms with high uncertainty avoidance need pre-dictability and have a strong tendency for the establishment of formal rules and process integration (Steensma et al. 2000). For example, as uncertainties in the supply chain increase, firms with high uncertainty avoidance tend to strengthen collaboration to share more information and leverage inventory, transportation, and planning to achieve certainty. To reduce uncertainty, firms tend to use elec-tronic linkages to augment interorganizational information processing capabilities to intensify communication and information sharing (Kim et al. 2005). In contrast, firms with low uncertainty avoidance value flexibility and tend to accept uncer-tainty and risk without uneasiness and tolerate various views and behaviors (Hofstede 2001; Wuyts and Geyskens 2005).

Based on organization theory, uncertainty has long been viewed as a dominant contingency (Thompson 1967; Bensaou and Venkatraman 1995) and is one of the underlying determinants of high transaction costs (Williamson 1975). There are many categories of uncertainty such as environmental, partnership, task, spe-cific capital assets, shared know-how, asymmetric information (e.g., holdup and information leakage), and complementary assets (Thompson 1967; Coase 1988; Fransman 1994; Bensaou and Venkatraman 1995; Dyer and Singh 1998; Kaufman et al. 2000). Uncertainty may present a firm with the need to renegotiate contracts and thus expose the firm to the risks of its partners' opportunism (Verwaal and Hesselmans 2004).

Reducing uncertainty via transparency and visibility of information flow is a major objective in supply chain collaboration (Holweg et al. 2005; Son et al. 2005). Market and technological uncertainty can effectively be dealt with through long-term partnerships in which supply chain partners share information of unexpected events and developments (Verwaal and Hesselmans 2004). The intense communication between supply chain partners also reduces behavioral uncertainty (e.g., opportun-ism) (Noordewier et al. 1990; Wuyts and Geyskens 2005). If there is no information sharing between partners, unpredictable or non-transparent demand patterns will cause demand amplification and bullwhip effect. This leads to poor service lev-els, high inventories, and frequent stock-outs (Forrester 1958; Sterman 1989; Lee et al. 1997). Thus, when facing high level of uncertainty, firms with uncertainty avoidance will tend to cooperate with supply chain partners in building collaborative inter-firm relationship.

## 3.4 Trust

Trust plays a major role in collaborative interorganizational relationship (Barney and Hansen 1994; Bromiley and Cummings 1995; Doney and Cannon 1997; Zaheer et al. 1998; Jarvenpaa and Tractinsky 1999; Pavlou 2002a, b; Johnson et al. 2004; Sheu et al. 2006; Du et al. 2011). Some view trust as the foundation of the digital market (Uzzi 1997; Keen 2000; Stewart and Segars 2002). From an economic view, trust leads to efficient transactions by reducing transaction costs (Bromiley and Cummings 1995). From a social exchange perspective, trust exists in the social context of supply chain partnerships creating social capital and affecting economic activities (Granovetter 1985; Uzzi 1997). In both views, trust has been regarded as a governance mechanism to reduce conflict and opportunism and promote cooperation, and further to enable firms to achieve collaborative advantage and better firm performance (Bradach and Eccles 1989; Barney and Hansen 1994; Morgan and Hunt 1994; Kumar et al. 1998; Zaheer et al. 1998).

Literature provides no unified definition of trust since its connotation is affected by the context attached to it (Palmer et al. 2000). Trust (i.e., intrerorganizational trust or partner trust) refers to the extent to which a firm subjectively believes that supply chain partners will perform work and transactions based on its confident expectations, regardless of its ability to check on their behaviors or monitor them (Gambetta 1988; Bhattacharya et al. 1998; Das and Teng 1998; Zaheer et al. 1998; Ba and Pavlou 2002; McKnight and Chervany 2002; Pavlou 2002a, b; Pavlou and Gefen 2004; Du et al. 2011). Trust refers to the degree to which a party has faith in another party's dependability and goodwill in an uncertain situation (Gambetta 1988; Ring and Van de Ven 1992; Nooteboom et al. 1997; Das and Teng 1998). It is the extent to which a party is willing to be vulnerable to another party's actions because it believes that the other party would not take advantage of an opportunity to gain at its expense given the chance (De Wever et al. 2005). Ba and Pavlou (2002) identify three sources of trust: familiarity (i.e., recurring exchanges that cause trust or mistrust); calculativeness (i.e., evaluation of the costs and benefits to the other's deceiving); and values (i.e., institutional measures that promote confidence in dependable behavior and goodwill).

Trust is one of the most accepted social standards for exchange coordination across organizations (Morgan and Hunt 1994; Jap 2001; Lejeune and Yakova 2005). It is a key relational attribute to build long-term relationships between supply chain partners as it motivates firms to tolerate short-term inequities in the belief that short-term inequities would be balanced out and compensated by mutual benefits over the long term (Son et al. 2005). Trust is also an informal mode of governance because it diminishes uncertainty in interorganizational exchange through self control (Koenig and van Wijk 1994; Kumar et al. 1998). The self control is demonstrated by replacing the calculative posture of risk-based judgments with favorable interpretations of another party's unmonitored activities (Uzzi 1997). The unspoken mutual anticipation and obligation produces an effective means of coordination (Kumar et al. 1998).

It has been reported that supply chain collaboration is difficult to implement because there has been an over reliance on technology and fundamentally a lack of trust between trading partners (Moberg and Speh 2003; Barratt 2004; Sheu et al. 2006). Trust is an important element for IOS enabled supply chain collaboration because trust can provide a foundation between collaborative partners for sharing critical information (Lejeune and Yakova 2005). However, trust between partners must be earned and trust comes only after the other party proves its abilities to offer solutions and also demonstrates loyalty (Min et al. 2005). Trust is achieved by behaving consistently over an extended period, e.g., maintaining quality standards without constant monitoring (Handfield and Nichols 1999; Lejeune and Yakova 2005).

Scholars agree that partner trust should be defined and measured as a multi-dimensional construct (Campbell 1992). Sako (1992) offers three dimensions of trust as contractual, competency and goodwill. Currall and Judge (1995) view trust as relationship activities, such as communication, informal agreement, absence of surveillance, and task coordination. Mayer et al. (1995) present three dimensions of trust: competence, integrity, and goodwill. McKnight and Chervany (2002) introduce four components of trust: competence, integrity, predictability, and benevolence. Johnson et al. (2004) identify two dimensions: dependability and benevolence. Despite diverse views, most trust definitions reflect two main elements: credibility and benevolence (Ring and Van de Ven 1992; Ganesan 1994; Doney and Cannon 1997; Johnson et al. 2004; Paul and McDaniel 2004; Sheu et al. 2006) (Table 3.4).

### 3.4.1 Credibility

Credibility refers to the extent to which a firm is confident about its supply chain partners' predictability, reliability, honesty, and competence (Pavlou 2002a; Johnson et al. 2004). This dimension corresponds to Johnson et al.'s (2004) dependability. It is the firm's expectation that supply chain partners will act in a dependable and predictable manner and can be counted on to perform their duties (Anderson and Weitz 1989). The firm will also hold a positive attitude toward the supply chain partner's honesty and integrity. For example, the company will believe that its partners will not share distorted information with it. The credibility dimension of trust denotes intentions of collaborative behaviors that may stem from making opportunism unreasonable or costly (Pavlou 2002a). Any long-term supply chain partnerships will require partners to fulfill their obligations and behave competently, consistently, and reliably (Zaheer et al. 1998; Tuten and Urban 2001).

### 3.4.2 Benevolence

Benevolence refers to the extent to which a firm expects that its supply chain partners will act fairly and will not take unfair advantage of the firm given the chance (Anderson and Narus 1990; Pavlou 2002a, b; Johnson et al. 2004). The

**Table 3.4** Definition of trust and subcomponents

| Construct | Definition | Literature |
|-----------|-----------|-----------|
| Trust | The extent to which a firm subjectively believes that supply chain partners will perform work and transactions based on its confident expectations, regardless of its ability to check on behavior or monitor them | Ba and Pavlou (2002), Bhattacharya et al. (1998), Das and Teng (1998), De Wever et al. (2005), Doney and Cannon (1997), Gambetta (1988), Johnson et al. (2004), McKnight and Chervany (2002), Nooteboom et al. (1997), Ring and Van de Ven (1992), Pavlou (2002a, b), Pavlou and Gefen (2004), Zaheer et al. (1998) |
| Credibility | The extent to which a firm is confident about its supply chain partners' predictability, reliability, honesty, and competence | Anderson and Weitz (1989), Johnson et al. (2004), Tuten and Urban (2001), Pavlou (2002a, b), Zaheer et al. (1998) |
| Benevolence | The extent to which a firm expects that its supply chain partners will act fairly and will not take unfair advantage of the firm given the chance | Anderson and Narus (1990); Baker et al. (1999), Borys and Jemison (1989), Ganesan (1994), Johnson et al. (2004), Pavlou (2002a, b), Sako (1992), Zaheer et al. (1998) |

benevolence dimension of trust is an expectation resulting from goodwill that firms will act fairly. Compared with credibility, benevolence is a higher level of trust because it is based on goodwill, not on rational calculation (Borys and Jemison 1989; Pavlou 2002a, b). The benevolence dimension represents true trust in that the firm believes that its partners would act in the firm's best interest even if there is no way of checking on or policing behavior (Sako 1992; Ganesan 1994; Zaheer et al. 1998; Baker et al. 1999; Johnson et al. 2004). It is the benevolence or goodwill component of trust that demonstrates trustworthiness, such as providing proprietary information or assistance without compensation (Johnson et al. 2004).

# References

Anderson, J. C., & Narus, J. A. (1990). A model of distributor firm and manufacturer firm working partnerships. *Journal of Marketing, 5*(41), 42–58.

Anderson, E., & Weitz, B. (1989). Determinants of continuity in conventional industrial channel dyads. *Marketing Science, 8*(4), 310–323.

Angeles, R., & Nath, R. (2001). Partner congruence in electronic data interchange (EDI) enabled relationships. *Journal of Business Logistics, 22*(2), 109–127.

Armstrong, C. P., & Sambamurthy, V. (1999). Information technology assimilation in firms: The influence of senior leadership and IT infrastructures. *Information Systems Research, 10*(4), 304–327.

Ashenbaum, B., Salzarulo, P., & Newman, W. (2012). Organizational structure, entrepreneurial orientation and trait preference in transportation brokerage firms. *Journal of Supply Chain Management, 48*(1), 3–23.

Attewell, P. (1992). Technology diffusion and organizational learning: The case of business computing. *Organization Science, 3*(1), 1–19.

Axelrod, R. (1984). *The Evolution of Cooperation*. New York: Basic Books.

Ba, S., & Pavlou, P. A. (2002). Evidence of the effect of trust building technology in electronic markets: Price premiums and buyer behavior. *MIS Quarterly, 26*(3), 243–268.

Bafoutsou, G., & Mentzas, G. (2002). Review and functional classification of collaborative systems. *International Journal of Information Management, 22*(4), 281–306.

Baker, T. L., Simpson, P. M., & Siguaw, J. A. (1999). The impact of suppliers' perceptions of reseller market orientation on key relationship constructs. *Journal of the Academy of Marketing Science, 27*(1), 50–57.

Barki, H., & Pinsonneault, A. (2005). A model of organizational integration, implementation effort, and performance. *Organization Science, 16*(2), 165–179.

Barney, J. B., & Hansen, M. H. (1994). Trustworthiness as a source of competitive advantage. *Strategic Management Journal, 15*(8), 175–190.

Barratt, M. (2004). Understanding the meaning of collaboration in the supply chain. *Supply Chain Management: An Internal Journal, 9*(1), 30–42.

Barret, S., & Konsynski, B. R. (1982). Interorganizational information sharing systems. *MIS Quarterly,* (Special Issue), 93–105.

Barua, A., Konana, P., Whinston, A. B., & Yin, F. (2004). An empirical investigation of net-enabled business value. *MIS Quarterly, 28*(4), 585–620.

Bates, K. A., Amundson, S. D., Schroeder, R. G., & Morris, W. T. (1995). The crucial interrelationship between manufacturing strategy and organizational culture. *Management Science, 41*(10), 1565–1580.

Bensaou, M., & Anderson, E. (1999). Buyer-supplier relations in industrial markets: When do buyers risk making idiosyncratic investments? *Organization Science, 10*(4), 460–481.

Bensaou, M., & Venkatraman, N. (1995). Configurations of interorganizational relationships: A comparison between U.S. and Japanese automakers. *Management Science, 41*(9), 1471–1492.

Bharadwaj, A. S. (2000). A resource based perspective on information technology capability and firm performance: An empirical investigation. *MIS Quarterly, 24*(1), 169–196.

Bhatt, G. D., & Grover, V. (2005). Types of information technology capabilities and their role in competitive advantage: An empirical study. *Journal of Management Information Systems, 22*(2), 253–277.

Bhattacharya, R., Devinney, T. M., & Pillutla, M. M. (1998). A formal model of trust based on outcomes. *Academy of Management Review, 23*(3), 459–472.

Boddy, D., Macbeth, D., & Wagner, B. (2000). Implementing collaboration between organizations: An empirical study of supply chain partnering. *Journal of Management Studies, 37*(7), 1003–1019.

Borys, B., & Jemison, D. (1989). Hybrid arrangements as strategic alliances: Theoretical issues in organizational combinations. *Academy of Management Review, 14*(2), 234–249.

Boynton, A. C., Zmud, R. W., & Jacobs, G. C. (1994). The influence of IT management practice on IT use in large organizations. *MIS Quarterly, 18*(3), 299–318.

Bradach, J. L., & Eccles, R. G. (1989). Price, authority, and trustTrust: From ideal types to plural forms. *Annual Review of Sociology, 15*(1), 97–118.

Broadbent, M., & Weill, P. (1997). Management by maxim: How business and IT managers can create IT infrastructures. *Sloan Management Review, 38*(3), 77–92.

Broadbent, M., Weill, P., & St. Clair, D. (1999). The implications of information technology infrastructure for business process redesign. *MIS Quarterly, 23*(2), 159–182.

Bromiley, P., & Cummings, L. L. (1995). Transaction costs in organizations with trust. In R. Bies, B. Sheppard, & R. Lewicki (Eds.), *Research on negotiation in organizations* (Vol. 5, pp. 219–250). Greenwich, CT: JAI Press.

Byrd, T. A., & Turner, D. E. (2000). Measuring the flexibility of information technology infrastructure: Exploratory analysis of a construct. *Journal of Management Information Systems, 17*(1), 167–208.

Cachon, G., & Lariviere, M. (2001). Contracting to assure supply: How to share demand forecasts in a supply chain. *Management Science, 47*(5), 629–646.

Campbell, A. (1992). The antecedents and outcomes of cooperative behaviors in international supply markets. *Unpublished Dissertation*. University of Toronto.

Caridi, M., Cigolini, R., & De Marco, D. (2005). Improving supply-chain collaboration by linking intelligent agents to CPFR. *International Journal of Production Research, 43*(20), 4191–4218.

Cash, J. I., & Konsynski, B. R. (1985). IS redraws competitive boundaries. *Harvard Business Review, 63*(2), 134–142.

Chang, H. L., & Shaw, M. J. (2004). *Developing the readiness index of IT-enabled supply chain collaboration*. Working Paper.

Chi, L., & Holsapple, C. W. (2005). Understanding computer-mediated interorganizational collaborationCollaboration: A model and framework. *Journal of Knowledge Management, 9*(1), 53–75.

Chin, W. W., Gopal, A., & Salisbury, W. D. (1997). Advancing the theory of adaptive structuration: The development of a scale to measure faithfulness of appropriation. *Information Systems Research, 8*(4), 342–367.

Christiaanse, E., & Venkatraman, N. (2002). Beyond SABRE: An empirical test of expertise exploitation in electronic channels. *MIS Quarterly, 26*(1), 15–38.

Chung, W., Chen, H., & Nunamaker, J. F., Jr. (2005). A visual framework for knowledge discovery on the web: An empirical study of business intelligence exploration. *Journal of Management Information Systems, 21*(4), 57–84.

Coase, R. H. (1988). *The firm, the market, and the law*. Chicago: University of Chicago Press.

Cohen, W. D., & Levinthal, D. A. (1990). Absorptive capacity: A new perspective on learning and innovation. *Administrative Science Quarterly, 35*(1), 128–152.

Collins, J., Youngdahl, B., Jamison, S., Mobasher, B., & Gini, M. (1998). A market architecture for multi-agent contracting. In K. Sycara & M. Wooldridge (Eds.), *Proceedings of the second international conference on autonomous agents* (pp. 285–292). Minneapolis, MN.

Currall, S. C., & Judge, T. A. (1995). Measuring trust between organizational boundary role persons. *Organizational Behavior Human Decision Processes, 64*(2), 151–170.

Das, T. K., & Teng, B. S. (1998). Between trust and control: Developing confidence in partner cooperation in alliances. *Academy of Management Review, 23*(3), 491–512.

Davenport, R., & Linder, J. (1994). Information management infrastructure: The new competitive weapon. *Proceedings of the 27th Annual Hawaii International Conference on Systems Sciences* (pp. 885–899). IEEE.

Dehning, B., & Richardson V. J. (2002). Returns on investments in information technology: A research synthesis. *Journal of Information Systems, 16*(1), 7–30.

De Wever, S., Martens, R., & Vandenbempt, K. (2005). The impact of trust on strategic resource acquisition through interorganizational networks: Towards a conceptual model. *Human Relations, 58*(12), 1523–1543.

DeSanctis, G., & Poole, M. S. (1994). Capturing the complexity in advanced technology use: Adaptive structuration theory. *Organization Science, 5*(2), 121–147.

Doney, P. M., & Cannon, J. P. (1997). An examination of the nature of trust in buyer-seller relationships. *Journal of Marketing, 61*(2), 35–51.

Droge, C., & Germain, R. (2000). The relationship of electronic data interchange with inventory and financial performance. *Journal of Business Logistics, 21*(2), 209–230.

Du, S., Bhattacharya, C., & Sankar, S. (2011). Corporate social responsibility and competitive advantage: Overcoming the trust barrier. *Management Science, 57*(9), 1528–1545.

Duncan, N. B. (1995). Capturing flexibility of information technology infrastructure: A study of resource characteristics and their measure. *Journal of Management Information Systems, 12*(2), 37–56.

Dyer, J. H. (1996). Does governance matter? Keiretsu alliances and asset specificity as sources of Japanese competitive advantage. *Organization Science, 7*(6), 649–666.

Dyer, J. H., & Singh, H. (1998). The relational view: Cooperative strategy and sources of interorganizational competitive advantage. *Academy of Management Review, 23*(4), 660–679.

El Sawy, O. A., Malhotra, A., Gosain, S. J., & Young, K. (1999). IT-intensive value innovation in the electronic economy: Insights from Marshall industries. *MIS Quarterly, 23*(3), 305–335.

Feeny, D. F., & Willcocks, L. P. (1998). Core IS capabilities for exploiting information technology. *Sloan Management Review, 39*(3), 9–21.

Fichman, R. G. (2000). The diffusion and assimilation of information technology innovations. In R. W. Zmud (Ed.), *Framing the domains of IT management: projecting the future through the past* (pp. 105–127). Cincinnati: Pinnaflex.

Forrester, J. W. (1958). Industrial dynamics: A major breakthrough for decision makers. *Harvard Business Review, 36*(4), 37–66.

Fransman, M. (1994). Information, knowledge, vision and theories of the firm. *Industrial and Corporate Change, 3*(3), 713–757.

Frohlich, M. T., & Westbrook, R. (2001). Arcs of integration: An international study of supply chain strategies. *Journal of Operations Management, 19*(2), 185–200.

Gambetta, D. (1988). *Trust: Making and breaking cooperative relations.* Oxford: Basil Blackwell.

Ganesan, S. (1994). Determinants of long-term orientation in buyer-seller relationships. *Journal of Marketing, 58*(2), 1–19.

Garcia-Dastugue, S. J., & Lambert, D. M. (2003). Internet-enabled coordination in the supply chain. *Industrial Marketing Management, 32*(3), 251–263.

Gini, M., & Boddy, M. (1998). Workshop on agent-based manufacturing, conducted at the *Autonomous Agents'98 Conference*, Minneapolis, MN.

Gopal, A., & Gosain, S. (2010). The role of organizational controls and boundary spanning in software development outsourcing: Implications for project performance. *Information Systems Research, 21*(4), 960–982.

Granovetter, M. S. (1985). Economic action and social structure: The problem of embeddedness. *American Journal Sociology, 91*(3), 481–510.

Gregory, K. L. (1983). Native-view paradigms multiple cultures and culture conflicts in organizations. *Administrative Science Quarterly, 28*, 359–376.

Grover, V., & Kohli, R. (2012). Cocreating IT value: New capabilities and metrics for multifirm environments. *MIS Quarterly, 36*(1), 225–232.

Grover, V., Teng, J., & Fiedler, K. (2002). Investigating the role of information technology in building buyer-supplier relationships. *Journal of Association for Information Systems, 3*, 217–245.

Gundlach, G. T., & Ernest, R. C. (1994). Exchange interdependence and interfirm interaction: Research in a simulated channel setting. *Journal of Marketing Research, 31*(4), 516–532.

Handfield, R. B., & Nichols, E. L., Jr. (1999). *Introduction to supply chain management.* Upper Saddle River: Prentice-Hall.

Hart, P., & Saunders, C. (1997). Power and trust: Critical factors in the adoption and use of electronic data interchange. *Organization Science, 8*(1), 23–43.

Hill, C. A., & Scudder, G. D. (2002). The use of electronic data interchange for supply chain coordination in the food industry. *Journal of Operations Management, 20*(4), 375–387.

Hofstede, G. (1980). *Culture's consequences: International differences in work-related values.* London: Sage Publications.

Hofstede, G. (1991). *Cultures and organizations: Software of the mind.* London: McGraw-Hill.

Hofstede, G. (1998). Identifying organizational subcultures: An empirical approach. *Journal of Management Studies, 35*(1), 1–12.

Hofstede, G. (2000). Masculine and feminine cultures. In A. E. Kazdin (Ed.), *Encyclopedia of psychology*, 5, Washington, DC: American Psychological Association.

Hofstede, G. (2001). *Culture's consequences* (2nd ed.). Thousand Oaks: Sage Publications.

Holweg, M., Disney, S., Holmström, J., & Småros, J. (2005). Supply chain collaboration: Making sense of the strategy continuum. *European Management Journal, 23*(2), 170–181.

Hong, I. B. (2002). A new framework for interorganizational systems based on the linkage of participants' roles. *Information & Management, 39*(4), 261–270.

Jap, S. D. (2001). Perspectives on joint competitive advantages in buyer–supplier relationships. *International Journal of Research in Marketing, 18*(1/2), 19–35.

Jarvenpaa, S. L., & Tractinsky, N. (1999). Consumer trust in an Internet store: A cross-cultural validation. *Journal of Computer-Mediated Communication, 5*(2), Online.

Johnson, D. A., McCutcheon, D. M., Stuart, F. I., & Kerwood, H. (2004). Effect of supplier trust on performance of cooperative supplier relationships. *Journal of Operations Management, 22*(1), 23–38.

Kambil, A., & Short, J. E. (1994). Electronic integration and business network redesign: A roles-linkage perspective. *Journal of Management Information Systems, 10*(4), 59–84.

Kaufman, A., Wood, C. H., & Theyel, G. (2000). Collaboration and technology linkages: A strategic supplier typology. *Strategic Management Journal, 21*(6), 649–663.

Keen, P. G. W. (1991). *Shaping the future: Business design through information technology.* Boston: Harvard Business School Press.

Keen, P.G.W. (2000). *Ensuring e-trust. Computerworld, 34* (11), 46–48.

Kekre, S., & Mukhopadhyay, T. (1992). Impacts of electronic data interchange on quality improvement and inventory reduction programs: A field study. *International Journal of Production Economics, 28*(3), 265–282.

Kim, K. K., Umanath, N. S., & Kim, B. H. (2005). An assessment of electronic information transfer in B2B supply-channel relationships. *Journal of Management Information Systems, 22*(3), 293–320.

King, W. R. (2002). IT capabilities, business processes, and impact on the bottom line. *Information Systems Management, 91*(2), 85–88.

Koenig, C., & van Wijk, G. (1994). Interorganizational collaboration: Beyond contracts. In *Workshop on schools of thought in strategic management. Beyond fragmentation?* ERASM/Erasmus University, Internal Report, December 14–15.

Kulp, S. C., Lee, H. L., & Ofek, E. (2004). Manufacturing benefits from information integration with retail customers. *Management Science, 50*(4), 431–444.

Kumar, K., & Van Dissel, H. G. (1996). Sustainable collaboration: Managing conflict and cooperation in interorganizational system. *MIS Quarterly, 20*(3), 279–300.

Kumar, K., Van Dissel, H. G., & Bielli, P. (1998). The merchant of Prato revisited: Toward a third rationality of information systems. *MIS Quarterly, 22*(2), 199–226.

Lee, H. L., Padmanabhan, V., & Whang, S. (1997). The bullwhip effect in supply chain. *Sloan Management Review, 38*(3), 93–102.

Lejeune, N., & Yakova, N. (2005). On characterizing the 4 C's in supply china management. *Journal of Operations Management, 23*(1), 81–100.

Malone, T. W., Yates, J., & Benjamin, R. I. (1987). Electronic markets and electronic hierarchies. *Communications of the ACM, 30*(6), 484–497.

Manthou, V., Vlachopoulou, M., & Folinas, D. (2004). Virtual e-Chain (VeC) model for supply chain collaboration. *International Journal of Production Economics, 87*(3), 241–250.

Massetti, B., & Zmud, R. W. (1996). Measuring the extent of EDI usage in complex organizations: Strategies and illustrative examples. *MIS Quarterly, 20*(3), 331–345.

Mata, F. J., Fuerst, W. L., & Barney, J. B. (1995). Information technology and sustained competitive advantage: A resource-based analysis. *MIS Quarterly, 19*(4), 487–505.

Mayer, R. C., Davis, J. H., & Schoorman, F. D. (1995). An integrative model of organizational trust. *Academy of Management Review, 20*(3), 709–734.

McAlister, L., Bazerman, M. H., & Fader, P. (1986). Power and goal setting in channel negotiations. *Journal of Marketing Research, 23*(3), 228–236.

McKeen, J. D., Smith, H. A., & Singh, S. (2005). Developments in practice: A framework for enhancing it capabilities. *Communications of the Association for Information Systems, 15,* 661–673.

McKenney, J. L., Copeland, D. C., & Mason, R. O. (1995). *Waves of change: Business evolution through information technology.* Boston: Harvard Business School Press.

McKnight, D. H., & Chervany, N. L. (2002). What trust means in e-commerce customer relationships: An interdisciplinary conceptual typology. *International Journal of Electronic Commerce, 6*(2), 35–53.

McLaren, T. S., Head, M. M., & Yuan, Y. (2004). Supply chain management information systems capabilities: An exploratory study of electronics manufacturers. *Information Systems and e-Business Management, 2,* 207–222.

Mehra, A., & Nissen, M. (1998). Case study: Intelligent software supply chain agents using ADE. In J. Baxter & B. Logan (Eds.), *Proceedings of the Workshop on Software Tools for Developing Agents* (pp. 53–62). Madison, WI: American Association for Artificial Intelligence.

Melville, N., Kraemer, K., & Gurbaxani, V. (2004). Information technology and organizational performance: An integrative model of it business value. *MIS Quarterly, 28*(2), 283–322.

Milton, N., Shadbolt, N., Cottam, H., & Hammersley, M. (1999). Towards a knowledge technology for knowledge management. *International Journal of Human-Computer Studies, 51*(3), 615–641.

Min, S., Roath, A., Daugherty, P. J., Genchev, S. E., Chen, H., & Arndt, A. D. (2005). Supply chain collaboration: What's happening? *International Journal of Logistics Management, 16*(2), 237–256.

Moberg, C. R., & Speh, T. W. (2003). Evaluating the relationship between questionable business practices and the strength of supply chain relationships. *Journal of Business Logistics, 24*(2), 1–19.

Morgan, R. M., & Hunt, S. D. (1994). The commitment-trust theory of relationship marketing. *Journal of Marketing, 58*(3), 20–38.

Mukhopadhyay, T., & Kekre, S. (2002). Strategic and operational benefits of electronic integration in B2B procurement process. *Management Science, 48*(10), 1301–1313.

Mukhopadhyay, T., Kekre, S., & Kalathur, S. (1995). Business value of information technology: A study of electronic data interchange. *MIS Quarterly, 19*(2), 137–157.

Narayandas, D., & Rangan, K. V. (2004). Building and sustaining buyer–seller relationships in mature industrial market. *Journal of Marketing, 68*(3), 63–77.

Nissen, M. E. (2000). Agent-based supply chain disintermediation vs. re-intermediation: Economic and technological perspectives. *International Journal of Intelligent Systems in Accounting, Finance and Management, 9*(4), 237–256.

Nissen, M. E., & Sengupta, K. (2006). Incorporating software agents into supply chains: Experimental investigation with a procurement task. *MIS Quarterly, 30*(1), 145–166.

Noordewier, T. G., George, J., & Nevin, J. R. (1990). Performance outcomes of purchasing arrangements in industrial buyer–partner relationships. *Journal of Marketing, 54*(4), 80–93.

Nooteboom, B., Berger, H., & Noorderhaven, N. G. (1997). Effects of trust and governance on relational risk. *Academy of Management Journal, 40*(2), 308–338.

O'Leary, D. E. (2003). Technologies for knowledge assimilation. In C. W. Holsapple (Ed.), *Handbook on Knowledge Management* (Vol. 2, pp. 29–46). New York: Springer.

Oliver, C. (1990). Determinants of interorganizational relationships: Integration and future directions. *Academy of Management Review, 15*(2), 241–265.

Orlikowski, W. J. (1993). Case tools as organizational change: Investigating incremental and radical changes in systems development. *MIS Quarterly, 17*(3), 309–340.

Palmer, J. W., Bailey, J. P., & Faraj, S. (2000). The role of intermediaries in the development of trust on the www: The use and effectiveness of trusted third parties and privacy statements. *Journal of Computer Mediated Communication, 5*(3), Online.

Paul, D. L., & McDaniel, R. R, Jr. (2004). A field study of the effect of interpersonal trust on virtual collaborative relationship performance. *MIS Quarterly, 28*(2), 183–227.

Pavlou, P. A. (2002a). Institution-based trust in interorganizational exchange relationships: The role of online B2B marketplaces on trust formation. *Journal of Strategic Information Systems, 11*(3/4), 215–243.

Pavlou, P. A. (2002b). Trustworthiness as a source of competitive advantage Competitive advantage in online auction markets. In D. H. Nagao (Ed.), *Best Paper Proceedings of the Academy of Management Conference* (pp. A1–A6) Denver, CO.

Pavlou, P. A., & Gefen, D. (2004). Building effective online marketplaces with institution-based trust. *Information Systems Research, 15*(1), 35–62.

Peppard, J., & Ward, J. (2004). Beyond strategic information systems: Towards an IS capability. *Journal of Strategic Information Systems, 13*(2), 167–194.

Piccoli, G., & Ives, B. (2005). IT-dependent strategic initiatives and sustained competitive advantage: A review and synthesis of the literature. *MIS Quarterly, 29*(4), 747–776.

Porter, M. E. (1980). *Competitive strategy*. New York: The Free Press.

Premkumar, G. P. (2000). Inter-organizational systems and supply chain management: An information processing perspective. *Information Systems Management, 17*(3), 57–69.

Premkumar, G., Ramamurthy, K., & Crum, M. R. (1997). Determinants of EDI adoption in the transportation industry. *European Journal of Information Systems, 6*(2), 107–121.

Rai, A., Pavlou, P., Im, G., & Du, S. (2012). Interfirm IT capability profiles and communications for cocreating relational value: evidence from the logistics industry. *MIS Quarterly, 36*(1), 233–267.

Ranganathan, C., Dhaliwal, J. S., & Teo, T. S. (2004). Assimilation and diffusion of web technologies in supply-chain management: An examination of key drivers and performance impacts. *International Journal of Electronic Commerce, 9*(1), 127–161.

Ravichandran, T., & Lertwongsatien, C. (2005). Effect of information systems resources and capabilities on firm performance: A resource-based perspective. *Journal of Management Information Systems, 21*(4), 237–276.

Ray, G., Muhanna, W. A., & Barney, J. B. (2005). Information technology and the performance of the customer service process: A resource-based analysis. *MIS Quarterly, 29*(4), 625–652.

Reichers, A. E., & Schneider, B. (1990). Climate and culture: An evolution of constructs. In B. Schneider (Ed.), *Organizational climate and culture*. San Francisco: Jossey-Bass Inc.

Ring, P. S., & Van de Ven, A. H. (1992). Structuring cooperative relationships between organizations. *Strategic Management Journal, 13*(7), 483–498.

Rockart, J. F. (1988). The line takes the leadership—IS management in a wired society. *Sloan Management Review, 29*(4), 55–64.

Rockart, J. F., Earl, M., & Ross, J. W. (1996). Eight imperatives for the new IT organization. *Sloan Management Review, 38*(1), 43–55.

Rodriguez-Aguilar, J. A., Martin, F. J., Noriega, P., Garcia, P., & Sierra, C. (1998). Competitive scenarios for heterogeneous trading agents. In K. Sycara & M. Wooldridge (Eds.), *Proceedings of the Second International Conference on Autonomous Agents* (pp. 293–300). Minneapolis, MN.

Ross, J. W., Beath, C. M., & Goodhue, D. L. (1996). Develop long-term competitiveness through IT assets. *Sloan Management Review, 38*(1), 31–42.

Saeed, K. A., Malhotra, M. K., & Grover, V. (2005). Examining the impact of interorganizational systems on process efficiency and sourcing leverage in buyer–supplier dyads. *Decision Sciences, 36*(3), 365–396.

Sako, M. (1992). *Prices, quality and trust: Inter-firm relations in Britain and Japan*. Cambridge: Cambridge University Press.

Sako, M., & Helper, S. (1998). Determinants of trust in supplier relations: Evidence from the automotive industry in Japan and the United States. *Journal of Economic Behavior & Organization, 34*(3), 387–417.

Salisbury, W. D., Chin, W. W., Gopal, A., & Newsted, P. R. (2002). Research report: Better theory through measurement—developing a scale to capture consensus on appropriation. *Information Systems Research, 13*(1), 91–103.

Sambamurthy, V., Bharadwaj, A., & Grover, V. (2003). Shaping agility through digital options: Reconceptualizing the role of information technology in contemporary firms. *MIS Quarterly, 27*(2), 237–263.

Santhanam, R., & Hartono, E. (2003). Issues in linking information technology capability to firm performance. *MIS Quarterly, 27*(1), 125–153.

Schein, E. H. (1985). *Organizational culture and leadership*. San Francisco: Jossey-Bass.

Schultze, U., & Orlikowski, W. J. (2004). A practice perspective on technology-mediated network relations: The use of Internet-based self-serve technologies. *Information Systems Research, 15*(1), 87–106.

Segil, L. (1998). Strategic alliances for the 21st century. *Strategy and Leadership, 26*(4), 12–16.

Sheu, C., Yen, H. R., & Chae, D. (2006). Determinants of supplier–retailer collaboration: Evidence from an international study. *International Journal of Operations & Production Management, 26*(1), 24–49.

Son, J., Narasimhan, S., & Riggins, F. J. (2005). Effects of relational factors and channel climate on EDI usage in the customer–supplier relationship. *Journal of Management information Systems, 22*(1), 321–353.

Steensma, H. K., Marino, L., Weaver, K. M., & Dickson, P. H. (2000). The influence of national culture on the formation of technology alliances by entrepreneurial firms. *Academy of Management Journal, 43*(5), 951–973.

Sterman, J. D. (1989). Modeling managerial behavior: Misperceptions of feedback in a dynamic decision making experiment. *Management Science, 35*(3), 321–339.

Stewart, K. A., & Segars, A. H. (2002). An empirical examination of the concern for information privacy instrument. *Information Systems Research, 13*(1), 36–49.

Subramani, M. (2004). How do suppliers benefit from information technology use in supply chain relationships? *MIS Quarterly, 28*(1), 45–73.

Subramaniam, C., & Shaw, M. J. (2002). A study of value and impact of B2B e-commerce: The case of web based procurement. *International Journal of Electronic Commerce, 6*(4), 19–40.

Teo, T., Nishant, R., Goh, M., & Agarwal, S. (2011). Leveraging collaborative technologies to build a knowledge sharing culture at HP Analytics. *MIS Quarterly Executive, 10*(1), 1–18.

Thomas, D., & Ranganathan, C. (2005). Enabling e-business transformation through alliances: Integrating social exchange and institutional perspectives. *Proceedings of the 38th Hawaii International Conference on System Sciences.*

Thompson, J. D. (1967). *Organizations in action.* New York: McGraw-Hill.

Tsui, E. (2003). Tracking the role and evolution of commercial knowledge management software. In C. W. Holsapple (Ed.), *Handbook on Knowledge Management* (Vol. 2, pp. 5–27). New York: Springer.

Tuten, T. L., & Urban, D. J. (2001). An expanded model of business-to-business partnership foundation and success. *Industrial Marketing Management, 30*(2), 149–164.

Uzzi, B. (1997). Social structure and competition in interfirm networks: The paradox of embeddedness. *Administrative Science Quarterly, 42*(1), 35–67.

Vegt, G. S., Jong, S. B., Bunderson, J. S., & Molleman, E. (2010). Power asymmetry and learning in teams: The moderating role of performance feedback. *Organization Science, 21*(2), 347–361.

Verwaal, E., & Hesselmans, M. (2004). Drivers of supply network governance: An explorative study industry. *European of the Dutch chemical Management Journal, 22*(4), 442–451.

Wade, M., & Hulland, J. (2004). The resource-based view and information systems research: Review, extension, and suggestions for future research. *MIS Quarterly, 23*(1), 107–142.

Wagner, J. A, I. I. I. (1995). Studies of individualism–collectivism: Effects on cooperation. *Academy of Management Journal, 38*(1), 152–172.

Walls, J. (1993). Global networking for local development: Task focus and relationship focus in cross-cultural communication. In L. M. Harasim (Ed.), *Global networks: Computers and international communication.* Cambridge: The MIT Press.

Weill, P. (1993). The role and value of information technology infrastructure: Some empirical observations. In R. Banker, R. Kauffman, & M. A. Mahmood (Eds.), *Strategic information technology management: Perspectives on organizational growth and competitive advantage.* Middleton: Idea Group Publishing.

Weill, P., Broadbent, M., & St. Clair, D. (1996). IT value and the role of IT infrastructure investments. In J. N. Luftman (Ed.), *Competing in the information age: Strategic alignment in practice* (pp. 361–384). New York: Oxford University Press.

Williamson, O. E. (1975). *Markets and hierarchies.* Englewood Cliffs: Prentice-Hall.

Wurman, P. R., Wellman, M. P., & Walsh, W. E. (1998). The Michigan internet AuctionBot: A configurable auction server for human and software agents. In K. Sycara & M. Wooldridge (Eds.), *Proceedings of the second international conference on autonomous agents* (pp. 301–308). Minneapolis, MN.

Wuyts, S., & Geyskens, I. (2005). The formation of buyer–supplier relationships: Detailed contract drafting and close partner selection. *Journal of Marketing, 69*(4), 103–117.

Zaheer, A., McEvily, B., & Perrone, V. (1998). Does trust matter? Exploring the effects of interorganizational and interpersonal trust on performance. *Organization Science, 9*(2), 141–159.

Zhu, K., & Kraemer, K. L. (2002). E-commerce metrics for net-enhanced organizations: Assessing the value of e-commerce to firm performance in the manufacturing sector. *Information Systems Research, 13*(3), 275–295.

# Chapter 4
# Supply Chain Collaboration Characterization

**Abstract** While individual success stories of supply chain collaboration have been reported, mainstream implementation has been much less successful than expected. Many collaborative relationships fail to meet the participants' expectations and few firms have truly capitalized on the potential of supply chain collaboration. In the literature, the exact nature and attributes of supply chain collaboration are not well understood and consistently defined. By highlighting the need for communication and joint knowledge creation as critical variables that are overlooked previously, we comprehensively define supply chain collaboration as seven interweaving components of information sharing, goal congruence, decision synchronization, incentive alignment, resources sharing, collaborative communication, and joint knowledge creation. In this chapter, we conceptualize supply chain collaboration by combining two streams of literature of process focus and relationship focus. We further define and elaborate each of the seven dimensions of supply chain collaborations.

In the face of information age and globalization, companies are increasingly emphasizing collaboration as a new source of competitive advantage (Dyer and Singh 1998). Supply chain collaboration has been strongly promoted by scholars and practitioners since the 1990s with some success stories of VMI, CFPR, and CR (Holweg et al. 2005). Despite its wide acceptance as an important issue, the concept for supply chain collaboration is not as well defined as it should be (Holweg et al. 2005; Simatupang and Sridharan 2005c; Nyaga et al. 2010; Fawcett et al. 2011; Allred et al. 2011; Verdecho et al. 2012; Fawcett et al. 2012). Supply chain collaboration has been defined in many different ways, and basically they fall into two groups of conceptualization: process focus and relationship focus.

First, supply chain collaboration has been viewed as a business process whereby two or more supply chain partners work together toward common goals and achieve more mutual benefits than can be achieved by acting alone (Mohr and Spekman 1994; Mentzer et al. 2001; Stank et al. 2001; Manthou et al. 2004; Sheu et al. 2006). The literature also reveals the importance of planning activities for

M. Cao and Q. Zhang, *Supply Chain Collaboration*,
DOI: 10.1007/978-1-4471-4591-2_4, © Springer-Verlag London 2013

collaborating among supply chain partners (Corbett et al. 1999; Narasimhan and Das 1999; Raghunathan 1999; Boddy et al. 2000; Ellinger 2000; Kaufman et al. 2000; Waller et al. 2000), integrating cross-functional processes (Lambert and Cooper 2000), coordinating the supply chain (Kim 2000), setting supply chain goals (Wong 1999; Peck and Juttner 2000), developing strategic alliances (McCutcheon and Stuart 2000; Whipple and Frankel 2000), establishing information-sharing parameters (Lamming et al. 2001), reviewing sourcing and outsourcing options (Ansari et al. 1999; Heriot and Kulkarni 2001), and defining supply chain power relationships among trading partners (Cox 1999; Maloni and Benton 2000; Cox 2001a, b, c; Watson 2001).

Second, supply chain collaboration has been portrayed as the formation of close, long term partnerships where supply chain members work jointly and share information, resources, and certain degrees of risk in order to accomplish mutual objectives (Sriram et al. 1992; Ellram and Edis 1996; Bowersox et al. 2003; Golicic et al. 2003). Firms "voluntarily agree to integrate human, financial, or technical resources in order to create a better business model" (Bowersox et al. 2003, p. 22).

There is evidence to suggest that partnerships are generally evolving phenomena (La Londe and Cooper 1989; Lundgren 1995) involving long term relationships between partners in the supply chain (Harland et al. 2004). Closeness has been widely identified as an important characteristic of relationships (Ellram 1991; Homburg 1995; Lambert et al. 1996; Saxton 1997; Macbeth 1998). Ellram and Hendrick (1995), pp. 41–42) define partnership as "an on-going relationship between two firms that involves a commitment over an extended time period, and a mutual sharing of information and the risks and rewards of the relationship". This definition is consistent with other descriptions in the literature that have defined supply chain partnerships as "relationships where customers and suppliers work together in a close, long-term relationship" (Burnes and New 1996) and "a situation in which there is an attempt to build close, long-term links between organizations in a supply chain that remain distinct, but which choose to work closely together" (Boddy et al. 2000). Kanter (1994) thinks the strongest and closest collaboration is supply chain partnership.

Many other definitions include the key aspects of common goals, joint activities, shared resources, shared risks/rewards, and trust (Dwyer et al. 1987; Gardner and Cooper 1988; Poirier and Houser 1993; Stuart and McCutcheon 1996; Brennan 1997; Skjoett-Larsen et al. 2003; Duffy and Fearne 2004; Plambeck et al. 2012). Poirier and Houser (1993, p. 56) describe the concept of partnering as "the creation of cooperative business alliances between an organization and its suppliers and customers. Business partnering occurs through a pooling of resources in a trusting atmosphere focused on continuous, mutual improvement". They argued that the greatest benefits of partnering are realized when all parties in the supply chain cooperate.

Ellram (1995) adds the most important dimension of information sharing, "an agreement between a buyer and a supplier that involves a commitment over an extended time period, and includes the sharing of information along with a sharing of the risks and rewards of the relationship." So does Macbeth (1998), "an approach to business in which companies expect a long-term relationship, develop complementary capabilities, share more information and engage in more joint planning than is customary. Sharing information during design may support more rapid product innovation".

Lambert et al.'s (2004, p.22) definition states: "A supply chain partnership is a tailored business relationship based on mutual trust, openness, shared risk and shared rewards that results in business performance greater than would be achieved by the two firms working together in the absence of partnership." The definition points out that the supply chain partnership is customized and incremental benefits must be gained from the tailoring effort, which consumes managerial time and talent (Lambert et al. 2004). Goffin et al. (2006) agree partnerships are not appropriate for the whole of the supplier base although this is almost universally assumed.

In addition, communication as a critical partnership variable should be emphasized. While research on communication within supply chain context is sparse, in the IOR and marketing channels literature, several academics have posed a link between communication and IOR governance structure (Mohr and Nevin 1990; Krapfel et al. 1991; Ring and Van de Ven 1992). To our knowledge, few studies have investigated communication in the supply chain (Olhager and Selldin 2003; Holden and O'Toole 2004; Prahinski and Benton 2004). Paralleling with Macneil's (1980) description of the differences in communication patterns between a discrete and a relational structure, Frazier et al. (1988) argue in a relational exchange, especially just-in-time relationships, communication would be frequent, both formal and informal, exchanging a considerable amount of information in connection with IOR processes as well as joint participation in long-term planning.

Another essential variable is partner-enabled knowledge creation (Malhotra et al. 2005). Shared or collective learning and knowledge creation is an important networking and collaborating activity. Powell et al. (1996) holds that supply chain collaboration offers a feasible means of obtaining intangible assets such as tacit knowledge and technological innovation. A supply chain with superior knowledge-transfer mechanism will be better able to compete on innovation (von Hippel 1988). By developing collaborative relations to suppliers rather than relying on arm's length relations, the rich flow of information should lead to improved learning, continuous improvement and better development solutions (Sako and Helper 1998).

Drawing on the literature, supply chain collaboration is defined as a long-term partnership process where supply chain partners work closely together to achieve common goals and mutual benefits. Specifically, it consists of seven components: information sharing, goal congruence, decision synchronization, incentive alignment, resources sharing, collaborative communication, and joint knowledge creation (Table 4.1). These seven components will be discussed in the following sections.

**Table 4.1** Definition of supply chain collaboration and subcomponents

| Construct | Definition | Literature |
|---|---|---|
| Supply chain collaboration | A long-term partnership process where supply chain partners work closely together to achieve common goals and mutual benefits. | Bafoutsou and Metzas 2002; Bowersox et al. 2003; Burnes and New 1996; Ellram and Hendrick 1995; Ellram and Edis 1996; Grieger 2003; Golicic et al. 2003; Johnson and Whang 2002; Kock and Nosek 2005; Lambert et al. 1996, 1999; Macbeth 1998; Manthou et al. 2004; Marquez et al. 2004; Mentzer et al. 2001; McDonnell 2001; Mohr and Nevin 1990; Poirier and Houser 1993; Sheu et al. 2006; Sriram et al. 1992; Stank et al. 2001; Nyaga et al. 2010; Allred et al. 2011; Fawcett et al. 2012 |
| Information sharing | The extent to which a firm shares a variety of relevant, accurate, complete and confidential information in a timely manner with its supply chain partners | Angeles and Nath 2003; Cooper, Ellram, Gardner, and Hanks 1997; Cooper, Lambert, and Pagh 1997; Kim and Umanath 2005; Monczka et al. 1998; Sheu et al. 2006; Simatupang and Sridharan 2005c, Stuart and McCutcheon 1996; Tyndall et al. 1998; Ren et al. 2010; Cheung et al. 2011 |
| Goal congruence | The extent to which supply chain partners perceive their own objectives to be satisfied by the accomplishment of the supply chain objectives | Angeles and Nath, 2001; Eliashberg and Michie 1984; Lejeune and Yakova 2005; Poirier and Houser 1993; Simatupang and Sridharan 2005a; Ryu and Yucesan 2010 |
| Decision synchronization | The process by which supply chain partners coordinate activities in supply chain planning and operations for optimizing the supply chain benefits | Corbett et al. 1999; Harland et al. 2004; Simatupang et al. 2002 |
| Incentive alignment | The process of sharing costs, risks, and benefits amongst supply chain partners | Clemons and Row 1992; Grandori and Soda1995; Melville et al. 2004; Sako 1992; Simatupang and Sridharan 2005b; Womack et al. 1990; Oliva and Watson 2011 |
| Resource sharing | The process of leveraging assets and making mutual asset investments amongst supply chain partners | Dwyer et al. 1987; Harland et al. 2004; Lambert et al. 1999; Simatupang et al. 2002; Simpson and Mayo 1997; Gomes and Dahab 2010 |

(continued)

**Table 4.1**  (continued)

| Construct | Definition | Literature |
|---|---|---|
| Collaborative communication | The contact and message transmission process among supply chain partners in terms of frequency, direction, mode, and influence strategy | Farace et al. 1977; Guetzkow 1965; Jablin, 1987; Mohr and Nevin 1990; Mohr et al. 1996; Prahinski and Benton 2004; Rogers and Agarwala-Rogers 1976 |
| Joint knowledge creation | The extent to which supply chain partners develop a better understanding of and response to the market and competitive environment by working together | Hardy et al. 2003; Johnson and Sohi 2003; Kaufman et al. 2000; Luo et al. 2006; Malhotra et al. 2005; Menon et al. 1999; Moorman 1995; Simonin 1997; Slater and Narver 1995; Srivastava et al. 1998; Gomes and Dahab 2010; Cheung et al. 2011 |

## 4.1  Information Sharing

Information sharing in the supply chain is critical and widely studied in the literature (Bowersox and Closs 1996; Walton 1996; Stock and Tatikonda 2000; Mentzer et al. 2001; Handfield and Bechtel 2002; Lejeune and Yakova 2005; Cheung et al. 2011; Ren et al. 2010). In the context of supply chain collaboration in particular, high levels of interdependence depend on high levels of information sharing (Boyacigiller 1990; Pahlberg 1997; Cannon and Perreault 1999; Bowersox et al. 2000; Kim et al. 2005). Information sharing is described as the "heart" (Lamming 1993, 1996), "lifeblood" (Stuart and McCutcheon 1996), "nerve center" (Chopra and Meindl 2001), "essential ingredient" (Min et al. 2005), "key requirement" (Sheu et al. 2006), and "foundation" (Lee and Whang 2001) of supply chain collaboration.

The Global Logistics Research Team at Michigan State University (1995) defines information sharing as the willingness to make strategic and tactical data such as inventory levels, forecasts, sales promotion, strategies, and marketing strategies available to firms forming supply chain nodes. Apart from exchange of demand information, exchange of more strategic information within a supply chain, including strategy, market, financial, technology, or new product information, may be important to ensure the long-term prosperity of partnerships (Liedtka 1996; Quinn 1999; Stank et al. 1999; Lee and Whang 2001; Harland et al. 2004; Simatupang and Sridharan 2004; Min et al. 2005; Nyaga et al. 2010; Allred et al. 2011; Fawcett et al. 2012). Uzzi (1997) argues that information shared in supply chain collaboration is more proprietary, tacit, and holistic than the transaction data (e.g., price and quantity) exchanged in arm's-length relationships. In line with Larson's (1992) results, it includes not only tacit information obtained through learning by doing but also data on profit margins and strategic information.

Ideally, supply chain partners can easily access real-time information online (Lee and Whang 2001; Manthou et al. 2004). The capability for all supply chain members to share timely information to complete transactions and to fulfill the requirements of shared business applications is called transparency of information (Angeles and Nath 2001), which is an effective way to counteract the problem of the bullwhip effect (i.e., demand information distortion in a supply chain). Advanced information and communications technologies (ICT), such as Internet-based EDI, may have great potential for improving information sharing to deal with the bullwhip effect and to enhance coordination across the entire supply chain (Scott-Morton 1991; Christopher 1992; Clemons and Row 1992; Harland et al. 2004; Kim et al. 2005; Simatupang and Sridharan 2005a). Thus, information transfer using ICT has the unique capability of simultaneously trimming both the firm's costs of decision and operation, and the transaction costs of its channel partner (Clemons and Row 1992). However, there is still little empirical research confirming the appropriate use of IT in information processing in the supply chain context (Harland et al. 2004).

Drawing on the literature, in current research, information sharing refers to the extent to which a firm shares a variety of relevant, accurate, complete and confidential information in a timely manner with its supply chain partners (Monczka et al. 1998; Angeles and Nath 2003; Simatupang and Sridharan 2005c, Sheu et al. 2006). Information sharing is generally conceptualized based on two dimensions: planning and monitoring supply chain operations (Stuart and McCutcheon 1996; Cooper, Ellram, et al. 1997; Cooper, Lambert et al. 1997; Tyndall et al. 1998; Angeles and Nath 2003; Kim and Umanath 2005; Simatupang and Sridharan 2005c). On the one hand, shared information provides a common base for partners and triggers the flows of products, services, funds, and feedback between the partners. On the other hand, shared information provides supply chain visibility that can trigger immediate, corrective actions relating to the flows of raw materials, finished goods, and services as needed (Min et al. 2005). Kim et al. (2005) view information sharing in a supply chain as the regulated flow of information from one unit (e.g., firm, work group, or individual) to the other unit.

Information sharing enables supply chain partners to see private data in another partner's systems and monitor the progress of products as they pass through each process in the supply chain. Thus, supply chain partners can make use of shared information to help fulfill demand more quickly with shorter order cycle times (Huang and Gangopadhyay 2004; Simatupang and Sridharan 2004; 2005b). Also, visibility of key performance metrics and process data enables the participating members to elicit the bigger picture of the situation that takes into account important factors in making effective decisions (Simatupang and Sridharan 2004). Effective decisions allow the chain members to address product flow issues more quickly, and thereby permit more agile demand planning to take place (Simatupang and Sridharan 2005a).

Several criteria, such as richness, frequency, depth, breadth, quality, speed, accuracy, timeliness, relevance, and reliability, can be employed to judge the contribution of information sharing to supply chain collaboration (Cannon and Homburg 2001; Mentzer et al. 2001; Rutner et al. 2001; Simpson et al. 2002; Simatupang and Sridharan 2004; Malhotra et al. 2005; Min et al. 2005; Sheu et al. 2006).

Data accuracy and timeliness are measured as the basis for improving the information sharing (Simatupang and Sridharan 2005a). In addition to sharing a broad range of information with partners, organizations should focus on improving the quality of information shared (Gosain et al. 2004). Handfield's (1993) instrument of information feedback is mainly composed of indicators such as information timeliness and volume of information. However, the study overlooks a vital component: the content of the information exchanged. It also does not address the medium's effect, i.e., the process utilized to transmit the information (Stuart and McCutcheon 1996).

In the following section, goal congruence, decision synchronization, incentive alignment, and resource sharing will be discussed. These four components are also collectively called process integration—the tight coupling of two or more processes through shared systems, automated functions and event triggers (e.g., auto replenishment) (Lockamy and McCormack 2004).

## 4.2 Goal Congruence

Angeles and Nath (2001) define goal congruence as the degree of goal agreement among supply chain partners. In the literature congruence is referred to as similarity, compatibility, or fit. Therefore, goal congruence between supply chain partners is the extent to which supply chain partners perceive their own objectives to be satisfied by the fulfillment of the supply chain objectives. There are two cases of true goal congruence: (1) supply chain partners believe that their objectives fully match those of the supply chain; (2) they believe that their objectives can be accomplished as an outcome of working toward the objectives of the supply chain (Lejeune and Yakova 2005; Ryu and Yucesan 2010). According to Eliashberg and Michie (1984), goal congruence refers to the degree of common goal accomplishment and it is used to assess the level of collaboration among supply chain partners.

The congruence concept presents the notion that supply chain collaboration need some degree of mutual understanding and agreement across certain organizational attributes, values, beliefs, and business practices. Goal congruence is regarded as a key element of supply chain partnership because it reduces the incentives for opportunism (Tjosvold 1986a, 1986b; Jap 2001; Naude and Buttle 2001). Several researchers have stressed the need for all partners in the collaborative relationship to clarify expectations carefully (Goffin et al. 2006). Supply chain partners should understand each other's goal and help each other accomplish the goal.

According to Poirier and Houser (1993, p. 201), "True supplier partnering requires an understanding of each party's needs and capabilities to establish a clear vision for focusing the efforts of people who work for buyer and supplier". In the last decade, top firms are developing extremely close relationships with selected clients and are placing significantly more emphasis on improved working arrangements with suppliers. The needs and capabilities of material suppliers, service suppliers, and especially customers are incorporated into strategic planning as firms view operations in terms of supply chain interactions and strategies (Stank et al. 2001). Inspired by collaborative goals, a firm is more willing to invest in and contribute to the development of supply chain partnership (Wong 1999).

Clear strategic goal leads to successful collaborative arrangements. It provides focus for the collaborative relationship and shapes interactions to gain the greatest cross-firm rewards/improvements. Without such a roadmap, optimal results cannot be achieved (Min et al. 2005). The importance of the strategic direction and the business vision of the participating firms are highlighted by Lambert et al. (1998). They argue that supply chain partners need to be in agreement about the supply chain management vision and key business processes underpinning this vision.

Landeros et al. (1995) think that expectations should be linked to performance measures. The mutual objective reflects the competitive factors that can be attained if the chain members build collaboration. Competitive factors can be in the form of product and service advantage, such as customer service, quality, price, supply chain costs, and responsiveness, recognized by the market as superior compared to competitors. These factors are assumed to enhance each chain member's profit, return-on-investment, and cash flow (Simatupang and Sridharan 2005a).

## 4.3 Decision Synchronization

Decision synchronization refers to the process by which supply chain partners coordinate activities in supply chain planning and operations for optimizing the supply chain benefits (Simatupang et al. 2002). Supply chain decisions include combining information and plans, resolving differences and conflicts, and establishing procedures, rules, and routines. Problems may arise in decision-making processes when information is widely dispersed or there is no clear authority structure. Decision-making mechanisms, which may incorporate routinized structures and procedures, can be developed through the coordination process (Harland et al. 2004). Whereas decision-making process has been the subject of many studies in organizational behavior research (March 1988), much less attention has been paid to it in supply chain research.

Planning decisions center on determining the efficient and effective way to use organizations' resources to achieve a specific set of objectives. There are seven key categories of supply chain planning decisions: operations strategy planning, demand management, production planning and scheduling, procurement, promise delivery, balancing change, and distribution management (Lockamy and McCormack 2004). Joint planning is required to align the operations and capacities of each collaborative partner. During the planning process, the manufacturer and its partners jointly prioritize goals and objectives based on individual company goal expectations (Min et al. 2005). Joint planning decisions may also include sales and order forecasts, customer service level, and pricing.

Joint operational decisions include inventory replenishment, order placement, order generation, and order delivery. Although supply chain partners synchronize their operational decisions, often the retailer has ultimate responsibility for the sales forecast and the supplier has ultimate responsibility for the order forecast and order generation. The interface team that is responsible for supporting this collaboration process consists of the retailer team (e.g. merchandising, purchasing, and distribution) and the supplier team (e.g. sales, planning/forecasting, and logistics) (Simatupang and Sridharan 2005a).

The difficulty of decision synchronization lies in the fact that supply chain partners have different decision rights and expertise about supply chain planning and operations (Simatupang and Sridharan 2005a). For example, a retailer may have the decision right to determine order quantity but not order delivery. Very often the supply chain partners have conflicting criteria in making decisions resulting in solutions that are less than optimum for the overall chain (Lee et al. 1997). The supply chain partners thus need to coordinate critical decisions that affect the way they achieve better performance. For example, VMI provides the supplier with decision rights to determine the frequency and quantity of orders that need to be delivered to the retailer's distribution center. This scheme enables the supplier to match supply with demand from the supply-chain-wide perspective and thereby improves profits for both members.

The way to judge the act of decision synchronization can be based on the responsiveness of the supply chain partners towards fulfilling customer demands and the effectiveness of joint decisions in enhancing supply chain profitability (Corbett et al. 1999). A level of synchronization in the decision-making process may be seen as a key element of collaboration in supply chain and as a way of building and maintaining a set of mutual partnerships (Harland et al. 2004). Information technology such as decision support system and virtual discussion forum can be used to implement decision synchronization effectively. For example, the use of an automated alert system in the exception cycle supports mutual response across the supply chain for satisfying customer demands (Simatupang and Sridharan 2004).

## 4.4 Incentive Alignment

Incentive alignment refers to the process of sharing costs, risks, and benefits amongst supply chain partners (Simatupang and Sridharan 2005b). It covers calculating costs, risks, and benefits as well as formulating incentive schemes. It is a critical factor to collaboration (Womack et al. 1990; Sako 1992; Clemons and Row 1992; Grandori and Soda 1995; Melville et al. 2004; Oliva and Watson 2011). Any successful supply chain management is based on close collaboration stimulated by mutual benefits (Lee and Whang 2001). The successful operation of supply chain partnerships mandates that each supply chain member should split gains and losses fairly and the collaboration outcome should be beneficial to all involved (Manthou et al. 2004).

Supply chain partners must align incentives for all members in order for collaboration to work. The incentive for each member should match its investment. Incentive alignment requires a detailed description of measures or procedures where the gains and risks are equitably allocated (Lee and Whang 2001). An appropriate incentive scheme can be devised in many different ways. Pay-for-effort is a scheme that links payment and effort. This assumes that rewarding effort would motivate the individual member to exert a given amount of effort that relates to a certain level of performance. Pay-for-performance is a scheme that links payment and performance. This scheme assumes that rewarding performance will motivate the individual chain member to achieve a particular level of performance. Equitable incentive is sharing the equitable load and benefits that result from exerting a certain amount of collaborative effort. The chain members accept the importance of the potential rewards that can be obtained from collaboration although costs need to be shared (Simatupang and Sridharan 2005b).

This scheme motivates the members to act in a manner consistent with the mutual strategic objectives such as making decisions that are optimal for the whole supply chain and revealing truthful private information (Simatupang and Sridharan 2005b). It secures sufficient levels of cooperation and commitment, while at the same time minimizing damaging routines such as opportunistic behavior. The practice whereby a customer acknowledges supplier achievement by granting awards is another way in which customers seek to motivate their suppliers. It may also involve the use of specific economic incentives, such as agreements to share future cost savings in component production costs. (Harland et al. 2004).

The contribution of incentive alignment can be judged based on compensation fairness and self-enforcement. Compensation fairness ensures that aligned incentives motivate the chain members to share equitably loads and benefits that result from collaborative efforts. An effective incentive scheme means that supply chain partners are self-enforcing for aligning their individual decisions with the mutual objective of improving total profits (Simatupang and Sridharan 2005b). Expert systems, activity-based costing, and Web-based technology can be used to trace, calculate, and display incentive scores (Kaplan and Narayanan 2001; Simatupang and Sridharan 2002).

## 4.5  Resource Sharing

Resource sharing refers to the process of leveraging assets and making mutual asset investments amongst supply chain partners. For example, a US manufacturer's international supplier can leverage the manufacturer's distribution networks with the other's market reach to distribute non-competitive products in the US market. This allows greater utilization of potentially slack resources (Min et al. 2005).

Resources leveraged include physical resources, such as manufacturing equipment, facility, and technology. Suppliers are often required to invest in manufacturing equipment that is dedicated to a particular customer; customers may also finance the equipment themselves which is then used by and within the supplier's plant (Harland et al. 2004). Facility configurations are observed in many Japanese networks, e.g. Toyota (Dyer 1996), leading to a closer-knit collaboration. The large body of literature on industrial clusters and regional networks discusses the importance of this phenomenon (e.g. Saxenian 1991). For example, many automotive suppliers re-locate and adapt their facilities to their large customers. Resources leveraged also include technologies. In the retailing sector, VMI or co-managed inventory (CMI) enable suppliers to assess stock-level data, via EDI, and take the necessary replenishment action (Scott-Morton 1991; Lamming 1996).

Sustainable collaborations must be supported with substantial mutual resource investments (Dwyer et al. 1987; Simpson and Mayo 1997; Gomes and Dahab 2010). Financial and non-financial investments including time, money, training, technology up-dates, and other resources are required. Reciprocal financial investment is usually present in an effective partnership (Lambert et al. 1999). The time and mutual effort required to achieve close relationships should not be underestimated (Goffin et al. 2006). Building and maintaining relationships and then dedicating personnel to managing the relationships, the processes, and the information are worth the effort. Collaborative relationships do not thrive unless they are encouraged and supported through sufficient commitment of management time (Min et al. 2005).

## 4.6 Collaborative Communication

An open communication mechanism is essential for companies engaged in the close interorganizational relationships such as supply chain partnership (Mohr et al. 1996; Stuart 1997; Tuten and Urban 2001; Holden and O'Toole 2004; Manthou et al. 2004; Goffin et al. 2006). Because the tight linkage between partners appears in different manners, communication channels must be well established and managed (Lee and Whang 2001).

Open, frequent, balanced, two-way, multilevel communications are generally thought to be an indication of a strong partnership (Carr and Pearson 1999; Lambert et al. 1999; Angeles and Nath 2001; Manthou et al. 2004). A more in-depth work done by Mohr and Nevin (1990) explore the pattern of communication from the mechanistic perspective of communication theory (Krone et al. 1987), in which communication is viewed as a transmission process through a channel (mode). Important facets of the communication process include the message (content), the channel (medium), feedback (bidirectional communication), and frequency (Guetzkow 1965; Rogers and Agarwala-Rogers 1976; Farace et al. 1977; Jablin 1987; Mohr and Nevin 1990). In line with Macneil (1980) and Frazier et al. (1988), Mohr and Nevin (1990) argue that communication patterns could be aligned along a continuum ranging from autonomous to collaborative and they coin the term "collaborative communication strategy" to refer to a particular combination of the facets of communication including higher frequency and more bidirectional flows, informal modes, and indirect content. This combination is likely to occur in channel conditions of relational structures, supportive climates, or symmetrical power.

Supply chain academicians have largely ignored the communication as a critical variable in supply chain collaboration. Holden and O'Toole (2004) examine if communication could delineate differing manufacturer–retailer relationships. Prahinski and Benton (2004) try to understand how suppliers think of their customers' evaluation on the communication process and determine its impact on suppliers' performance. Several other studies assess the indirect influence strategy (communication content) or formality (communication medium) on the buying firm's performance (Srinivasan et al. 1994; Walton and Marucheck 1997; D'Amours et al. 1999; Krause et al. 2000).

As in Mohr et al. (1996), collaborative communication is defined in this research as the contact and message transmission process among supply chain partners in terms of frequency, direction, mode, and influence strategy. Supply chain partners tend to establish communication based on higher frequency, more bidirectional flows, informal modes, and indirect influence strategy. Collaborative communication in supply chain can serve as the channel by which information is shared, goal is matched, decision making is synchronized, incentive is aligned, resources is coordinated, and joint knowledge is created.

Frequency refers to the amount of contact between supply chain partners to conduct supply chain activities adequately (Farace et al. 1977; Mohr et al. 1996). In evaluating the frequency of communication, one should examine the amount of contact in relation to the amount of contact necessary to conduct activities adequately because too much contact can overload supply chain members and have dysfunctional consequences (Guetzkow 1965; Mohr et al. 1996).

Direction refers to the movement of communication between supply chain partners. Bi-directionality means two-way movement (both upward and downward) of communication along the supply chain (Purdy et al. 1994; Mohr and Sohi 1995; Prahinski and Benton 2004). Unidirectional communication flows (upward or downward) would hold only if one member in the supply chain is more powerful (Mohr and Nevin 1990).

Mode, also called medium, refers to the method used by supply chain partners to transmit information. Two major classification schemes are: medium richness and formality. Medium richness is the number of cues that can be used by the receiver to interpret the message (Daft and Lengel 1986). The authors identify medium richness in descending order as follows: face-to-face meetings, telephone, letters and memos, impersonal documents and numeric documents. Formality assesses the structure and routine of the communication (Carr and Pearson 1999; Mohr and Sohi 1995). Because of the categorical nature of medium richness, communication formality will be studied in this research. While formal mode refers to the communication established through structured rules and fixed procedures, informal mode is defined as the degree to which the communication between supply chain partners is established through spontaneous and non-regularized manner, such as word-of-mouth contacts.

Influence strategy of communication is embedded in the communication content (i.e. the message that is transmitted). Using direct influence strategies, a firm tries to change behaviors of its supply chain partners by implying or requesting the specific action that the firm wants its partners to take. Examples of direct influence strategies include requests, recommendations, promises, and appeals to legal obligations. Indirect influence strategies are designed to change the supply chain partners' beliefs and attitudes about the desirability of the intended behavior; no specific action is requested directly. An example of indirect influence strategies is information exchange, whereby the firm uses discussions on general business issues and operating procedures to alter its partner's attitude about desirable behaviors (Frazier and Summers 1984; Mohr and Nevin 1990).

Because supply chain partners need to share more information in order to coordination more closely shared activities, a higher level of communication frequency may be necessary (Huber and Daft 1987). For better coordination of activities, communication will flow both upward and downward in the supply chain structures (Dwyer et al. 1987). Because supply chain partners are closely linked, communication among them is generally more informal. Though formal communication modes are also used, the tighter linkages between supply chain partners allow for more informal interactions (Mohr and Nevin 1990). Because supply chain partners are more willing to share benefits and risks, simply providing information to other members may be sufficient to encourage them to play a part. Thus influence strategies are more indirect than direct. Also, interdependent partners tend not to use of tough, distributive bargaining tactics (Stohl and Redding 1987).

## 4.7 Joint Knowledge Creation

Joint knowledge creation refers to the extent to which supply chain partners develop a better understanding of and response to the market and competitive environment by working together (Malhotra et al. 2005). While collaboration facilitates information sharing, joint knowledge creation is one of the primary objectives of collaboration (Simonin 1997; Hardy et al. 2003; Gomes and Dahab 2010; Cheung et al. 2011). There are two kinds of knowledge creation activities: knowledge exploration (i.e., search and acquire new and relevant knowledge) and knowledge exploitation (i.e., assimilate and apply relevant knowledge) (Bhatt and Grover 2005). The capture, exchange, and assimilation of knowledge (e.g., process, technology, or market knowledge) between supply chain partners enable innovation and the long-term competitiveness of the supply chain as a whole (Harland et al. 2004).

Supply chain collaboration stimulates collective learning for improving supply chain performance as a whole that brings benefits to all participating members (Simatupang and Sridharan 2004). Supply chain partners should engage in building the knowledge base together, and more importantly, involve dissemination and shared interpretation that enable firms to create new values such as developing new products, building brand image, responding to customers' needs, and establishing channel relationships (Menon et al. 1999; Moorman 1995; Srivastava et al. 1998; Johnson and Sohi 2003; Slater and Narver 1995; Luo et al. 2006; Kaufman et al. 2000).

It has been demonstrated that markets are not effective structures to access and transfer intangible, tacit assets, e.g., knowledge (Barney 1991; Sobrero and Roberts 2001). Supply chain collaborations provide a way of exchanging tacit knowledge by establishing direct links with knowledge sources or engaging in joint development (Roberts and Berry 1985; Lorenzoni and Baden-Fuller 1995). Recent research confirms that the strategic value of supply chain collaborative arrangement is not only to increase efficiency, but also to assimilate external knowledge (Clark 1989; Dyer 1997; Sobrero and Roberts. 2001). Partnering is very useful for companies to follow the latest trends, and through partnering companies can achieve a time advantage over competitors by obtaining information from both suppliers and customers (Verwaal and Hesselmans 2004).

# References

Allred, C. R., Fawcett, S. E., Wallin, C., & Magnan, G. M. (2011). A dynamic collaboration capability as a source of competitive advantage. *Decision Sciences, 42*(1), 129–161.

Angeles, R., & Nath, R. (2001). Partner congruence in electronic data interchange (EDI) enabled relationships. *Journal of Business Logistics, 22*(2), 109–127.

Angeles, R., & Nath, R. (2003). Electronic supply chain partnerships: Reconsidering relationship attributes in customer-supplier dyads. *Information Resources Management Journal, 16*(3), 59–84.

Ansari, A., Lockwood, D. L., & Modarress, B. (1999). Supplier product integration: A new competitive approach. *Production and Inventory Management Journal, 40*(3), 57–61.

Bafoutsou, G., & Mentzas, G. (2002). Review and functional classification of collaborative systems. *International Journal of Information Management, 22*(4), 281–306.

Barney, J. (1991). Firm resources and sustained competitive advantage. *Journal of Management, 17*(1), 99–120.

Bhatt, G. D., & Grover, V. (2005). Types of information technology capabilities and their role in competitive advantage: An empirical study. *Journal of Management Information Systems, 22*(2), 253–277.

Boddy, D., Macbeth, D., & Wagner, B. (2000). Implementing collaboration between organizations: An empirical study of supply chain partnering. *Journal of Management Studies, 37*(7), 1003–1019.

Bowersox, D. J., & Closs, D. C. (1996). *Logistical management: The integrated supply chain process*. New York: McGraw-Hill.

Bowersox, D. J., Closs, D. J., & Stank, T. P. (2000). Ten mega-trends that will revolutionize supply chain logistics. *Journal of Business Logistics, 21*(2), 1–16.

Bowersox, D. J., Closs, D. J., & Stank, T. P. (2003). How to master cross-enterprise collaboration. *Supply Chain Management Review, 7*(4), 18–27.

Boyacigiller, N. (1990). The role of expatriates in the management of interdependence, complexity, and risk in multinational corporations. *Journal of International Business Studies, 21*(3), 357–381.

Brennan, R. (1997). Buyer/supplier partnering in British industry: The automotive and telecommunications sectors. *Journal of Marketing Management, 13*(8), 759–775.

Burnes, B., & New, S. (1996). Understanding supply chain improvement. *European Journal of Purchasing and Supply Management, 2*(1), 21–30.

Cannon, J., & Homburg, C. (2001). Buyers-supplier relationships and customer firm costs. *Journal of Marketing, 65*(1), 29–43.

Cannon, J. P., & Perreault, W. D. (1999). Buyer-seller relationships in business markets. *Journal of Marketing Research, 36*(4), 439–460.

Carr, A. S., & Pearson, J. N. (1999). Strategically managed buyer–supplier relationships and performance outcomes. *Journal of Operations Management, 17*(5), 497–519.

Cheung, M., Myers, M., & Mentzer, J. (2011). The value of relational learning in global buyer-supplier exchanges: a dyadic perspective and test of the pie-sharing premise. *Strategic Management Journal, 32*, 1061–1082.

Chopra, S., & Meindl, P. (2001). *Supply chain management: Strategy, planning and operation.* Upper Saddle River: Prentice-Hall.

Christopher, M. G. (1992). *Logistics and supply chain management.* London: Pitman.

Clark, K. B. (1989). The interaction of design hierarchies and market concepts in technological evolution. *Research Policy, 14*, 235–251.

Clemons, E., & Row, M. (1992). Information technology and industrial cooperation: The changing economics of coordination and ownership. *Journal of Management Information Systems, 9*(2), 9–28.

Cooper, M., Ellram, L. M., Gardner, J. T., & Hanks, A. M. (1997a). Meshing multiple alliances. *Journal of Business Logistics, 18*(1), 67–89.

Cooper, C., Lambert, C., & Pagh, D. (1997b). Supply chain management: More than a new name for logistics. *The International Journal of Logistics Management, 8*(1), 1–14.

Corbett, C. J., Blackburn, J. D., & Van Wassenhove, L. N. (1999). Partnerships to improve supply chains. *Sloan Management Review, 40*(4), 71–82.

Cox, A. (1999). Power, value and supply chain management. *Supply Chain Management, 4*(4), 167–175.

Cox, A. (2001a). The power perspective in procurement and supply management. *Journal of Supply Chain Management, 37*(2), 4–7.

Cox, A. (2001b). Understanding buyer and supplier power: A framework for procurement and supply competence. *Journal of Supply Chain Management, 37*(2), 8–15.

Cox, A. (2001c). Managing with power: Strategies for improving value appropriation from supply relationships. *Journal of Supply Chain Management, 37*(2), 42–47.

D'Amours, S., Montreuil, B., Lefrancois, P., & Soumis, F. (1999). Networked manufacturing: the impact of information sharing. *International Journal of Production and Economics, 58*(1), 63–79.

Daft, R. L., & Lengel, R. H. (1986). Organizational information requirements, media richness, and structural design. *Management Science, 32*(5), 554–571.

Duffy, R., & Fearne, A. (2004). The impact of supply chain partnerships on supplier performance. *International Journal of Logistics Management, 15*(1), 57–71.

Dwyer, F. R., Schurr, P. H., & Oh, S. (1987). Developing buyer-seller relationships. *Journal of Marketing, 51*(2), 11–27.

Dyer, J. H. (1996). Does governance matter? Keiretsu alliances and asset specificity as sources of Japanese competitive advantage. *Organization Science, 7*(6), 649–666.

Dyer, J. H. (1997). Effective interfirm collaboration: How firms minimize transaction costs and maximize transaction value. *Strategic Management Journal, 18*(7), 535–556.

Dyer, J. H., & Singh, H. (1998). The relational view: Cooperative strategy and sources of interorganizational competitive advantage. *Academy of Management Review, 23*(4), 660–679.

Eliashberg, J., & Michie, D. A. (1984). Interorganizational competitive advantage. *Academy of Management Review, 23*(4), 660–679.

Ellinger, A. E. (2000). Improving marketing/logistics cross-functional collaborations in the supply chain. *Industrial Marketing Management, 29*(1), 85–96.

Ellram, L. M. (1991). A managerial guideline for the development and implementation of purchasing partnerships. *International Journal of Purchasing and Materials Management, 27*(3), 2–8.

Ellram, L. M. (1995). A managerial guideline for the development and implementation of purchasing partnerships. *International Journal of Purchasing and Materials Management, 31*(2), 9–16.

Ellram, L. M., & Edis, O. R. V. (1996). A case study of successful partnering implementation. *International Journal of Purchasing and Materials Management, 32*(2), 20–28.

Ellram, L. M., & Hendrick, T. E. (1995). Partnering characteristics: A dyadic perspective. *Journal of Business Logistics, 16*(1), 41–64.

Farace, R., Monge, P., & Russell, H. (1977). *Communicating and organizing reading.* MA: Addison-Wesley.

Fawcett, S. E., Fawcett, A., Watson, B., & Magnan, G. (2012). Peeking inside the black box: toward an understanding of supply chain collaboration dynamics. *Journal of Supply Chain Management, 48*(1), 44–72.

Fawcett, S. E., Wallin, C., Allred, C., Fawcett, A., & Magnan, G. M. (2011). Information technology as an enabler of supply chain collaboration: A dynamic-capabilities perspective. *Journal of Supply Chain Management, 47*(1), 38–59.

Frazier, G. L., Spekman, R. E., & O'Neal, C. R. (1988). Just-in-time exchange relationships in industrial markets. *Journal of Marketing, 52*(4), 52–67.

Frazier, G., & Summers, J. (1984). Interfirm influence strategies and their application within distribution channels. *Journal of Marketing, 48*(3), 43–55.

Gardner, J., & Cooper, M. C. (1988). Elements of strategic partnership. In J. E. McKeon (Ed.), *Partnerships: A natural evolution in logistics* (pp. 15–32). Cleveland: Logistics Research, Inc.

Global Logistics Research Team at Michigan State University. (1995). *World class logistics: The challenge of managing continuous change.* Oak Brook: Council of Logistics Management.

Goffin, K., Lemke, F., & Szwejczewski, M. (2006). An exploratory study of close supplier-manufacturer relationships. *Journal of Operations Management, 24*(2), 189–209.

Golicic, S. L., Foggin, J. H., & Mentzer, J. T. (2003). Relationship magnitude and its role in interorganizational relationship structure. *Journal of Business Logistics, 24*(1), 57–75.

Gomes, P., & Dahab, S. (2010). Bundling resources across supply chain dyads: The role of modularity and coordination capabilities. *International Journal of Operations & Production Management, 30*(1), 57–74.

Gosain, S., Malhotra, A., & El Sawy, O. A. (2004). Coordinating for flexibility in e-business supply chains. *Journal of Management Information Systems, 21*(3), 7–45.

Grandori, A., & Soda, G. (1995). Inter-firm networks: Antecedents mechanisms and forms. *Organization Studies, 16*(2), 183–214.

Grieger, N. (2003). Electronic marketplaces: A literature review and a call for supply chain management research. *European Journal of Operational Research, 144*(2), 280–294.

Guetzkow, H. (1965). Communications in organizations. In J. March (Ed.), *Handbook of Organizations* (pp. 534–573). Chicago: Rand McNally.

Handfield, R. B. (1993). A resource dependence perspective of just-in-time purchasing. *Journal of Operations Management, 11*(3), 289–311.

Handfield, R. B., & Bechtel, C. (2002). The role of trust and relationship structure in improving supply chain responsiveness. *Industrial Marketing Management, 31*(4), 367–382.

Hardy, C., Phillips, N., & Lawrence, T. B. (2003). Resources, knowledge and influence: The organizational effects of interorganizational collaboration. *Journal of Management Studies, 40*(2), 321–347.

Harland, C. M., Zheng, J., Johnsen, T. E., & Lamming, R. C. (2004). A conceptual model for researching the creation and operation of supply networks. *British Journal of Management, 15*(1), 1–21.

Heriot, K. C., & Kulkarni, S. P. (2001). The use of intermediate sourcing strategies. *Journal of Supply Chain Management, 37*(1), 18–26.

Holden, M. T., & O'Toole, T. (2004). A quantitative exploration of communication's role in determining the governance of manufacturer–retailer relationships. *Industrial Marketing Management, 33*(6), 539–548.

Holweg, M., Disney, S., Holmström, J., & Småros, J. (2005). Supply chain collaboration: Making sense of the strategy continuum. *European Management Journal, 23*(2), 170–181.

Homburg, C. (1995). Closeness to the customer in industrial markets: towards a theory-based understanding of measurement, organizational antecedents, and performance outcomes. *Zeitschrift für Betriebswirtschaft, 65*(3), 309–331.

Huang, Z., & Gangopadhyay, A. (2004). A simulation study of supply chain management to measure the impact of information sharing. *Information Resources Management Journal, 17*(3), 20–31.

Huber, G., & Daft, R. (1987). The information environment of organization. In F. Jablin, et al. (Eds.), *Handbook of organizational communication: An interdisciplinary perspective* (pp. 130–164). Newbury Park: Sage Publications, Inc.

Jablin, F. (1987). Formal organization structure. In F. Jablin, et al. (Eds.), *Handbook of organizational communication: An interdisciplinary perspective* (pp. 389–419). Newbury Park: Sage Publications, Inc.

Jap, S. D. (2001). Perspectives on joint competitive advantages in buyer-supplier relationships. *International Journal of Research in Marketing, 18*(1/2), 19–35.

Johnson, M. E., & Whang, S. (2002). E-business and supply chain management: An overview and framework. *Production and Operations Management, 11*(4), 413–422.

Johnson, J. J., & Sohi, R. S. (2003). The development of interfirm partnering competence: Platforms for learning, learning activities and consequences of learning. *Journal of Business Research, 56*(9), 757–766.

Kanter, R.M. (1994). Collaborative advantage: The art of alliances. *Harvard Business Review,* 96–108.

Kaplan, R., & Narayanan, V. G. (2001). Measuring and managing customer profitability. *Journal of Cost Management, 15*(5), 5–15.

Kaufman, A., Wood, C. H., & Theyel, G. (2000). Collaboration and technology linkages: A strategic supplier typology. *Strategic Management Journal, 21*(6), 649–663.

Kim, B. (2000). Coordinating an innovation in supply chain management. *European Journal of Operational Research, 123*(3), 568–584.

Kim, K. K., & Umanath, N. S. (2005). Information transfer in B2B procurement: An empirical analysis and measurement. *Information and Management, 42*(6), 813–828.

Kim, K. K., Umanath, N. S., & Kim, B. H. (2005). An assessment of electronic information transfer in B2B supply-channel relationships. *Journal of Management Information Systems, 22*(3), 293–320.

Kock, N., & Nosek, J. (2005). Expanding the boundaries of e-collaboration. *IEEE Transactions on Professional Communication, 48*(1), 1–9.

Krapfel, R. E., Salmond, D., & Spekman, R. (1991). A strategic approach to managing buyer–seller relationships. *European Journal of Marketing, 25*(9), 22–37.

Krause, D. R., Scannell, T. V., & Calantone, R. J. (2000). A structural analysis of the effectiveness of buying firm's strategies to improve supplier performance. *Decision Sciences, 31*(1), 33–55.

Krone, K., Jablin, F., & Putnam, L. (1987). Communication theory and organizational communication: Multiple perspectives. In F. Jablin, et al. (Eds.), *Handbook of organizational communication: An interdisciplinary perspective* (pp. 11–17). Newbury Park: Sage Publications, Inc.

La Londe, B. J., & Cooper, M. C. (1989). *Partnerships in providing customer services: A third-party perspective.* Oak Brook: Council of Logistics Management.

Lambert, D. M., & Cooper, M. C. (2000). Issues in supply chain management. *Industrial Marketing Management, 29*(1), 65–83.

Lambert, D. M., Cooper, M. C., & Pugh, J. D. (1998). Supply chain management: Implementation issues and research opportunities. *International Journal of Logistics Management, 9*(2), 1–19.

Lambert, D. M., Emmelhainz, M. A., & Gardner, J. T. (1996). Developing and implementing supply chain partnerships. *The International Journal of Logistics Management, 7*(2), 1–17.

Lambert, D. M., Emmelhainz, M. A., & Gardner, J. T. (1999). Building successful logistics partnerships. *Journal of Business Logistics, 20*(1), 165–182.

Lambert, D. M., Emmelhainz, M. A., & Gardner, J. T. (2004). Supply chain partnerships: Model validation and implementation. *Journal of Business Logistics, 25*(2), 21–42.

Lamming, R. C. (1993). *Beyond partnership: Strategies for innovation and lean supply*. Hemel Hempstead: Prentice Hall.

Lamming, R. C. (1996). Squaring lean supply with supply chain management. *International Journal of Operations & Production Management, 10*(2), 183–196.

Lamming, R. C., Caldwell, N. D., Harrison, D. A., & Phillips, W. (2001). Transparency in supply relationships: Concept and practice. *Journal of Supply Chain Management, 37*(4), 4–10.

Landeros, R., Reck, R., & Plank, R. E. (1995). Maintaining buyer–supplier partnerships. *International Journal of Purchasing and Materials Management, 31*(3), 3–11.

Larson, A. (1992). Network dyads in entrepreneurial settings: A study of the governance of exchange processes. *Administrative Science Quarterly, 37*(1), 76–104.

Lee, H. L., Padmanabhan, V., & Whang, S. (1997). The bullwhip effect in supply chain. *Sloan Management Review, 38*(3), 93–102.

Lee, H.L., & Whang, S. (2001). E-Business and supply chain integration. *Stanford Global Supply Chain Management Forum*, SGSCMF-W2-2001.

Lejeune, N., & Yakova, N. (2005). On characterizing the 4 C's in supply china management. *Journal of Operations Management, 23*(1), 81–100.

Liedtka, J. M. (1996). Collaborating across lines of business for competitive advantage. *Academy of Management Executive, 10*(2), 20–37.

Lockamy, A., & McCormack, K. (2004). Linking SCOR planning practices to supply chain performance: An exploratory study. *International Journal of Operations & Production Management, 24*(12), 1192–1218.

Lorenzoni, G., & Baden-Fuller, C. (1995). Creating a strategic center to manage a web of partners. *California Management Review, 37*(3), 146–162.

Lundgren, A. (1995). *Technological innovation and network evolution*. London: Routledge.

Luo, X., Slotegraaf, R. J., & Pan, X. (2006). Cross-functional coopetition: The simultaneous role of cooperation and competition within firms. *Journal of Marketing, 70*(2), 67–80.

Macbeth, D. (1998). Partnering—why not? In *The Second Worldwide Research Symposium on Purchasing and Supply Chain Management*, (pp. 351–362) England: Stamford.

Macneil, I. R. (1980). *The new social contract: An inquiry into modern contractual relations*. New Haven: Yale University Press.

Malhotra, A., Gasain, S., & El Sawy, O. A. (2005). Absorptive capacity configurations in supply chains: Gearing for partner-enabled market knowledge creation. *MIS Quarterly, 29*(1), 145–187.

Maloni, M., & Benton, W. C. (2000). Power influences in the supply chain. *Journal of Business Logistics, 21*(1), 49–73.

Manthou, V., Vlachopoulou, M., & Folinas, D. (2004). Virtual e-Chain (VeC) model for supply chain collaboration. *International Journal of Production Economics, 87*(3), 241–250.

March, J. G. (1988). *Decisions and organizations*. Oxford: Basil Blackwell.

Marquez, A. C., Bianchi, C., & Gupta, J. N. D. (2004). Operational and financial effectiveness of e-collaboration tools in supply chain integration. *European Journal of Operational Research, 159*(2), 348–363.

McCutcheon, D., & Stuart, F. I. (2000). Issues in the choice of supplier alliance partners. *Journal of Operations Management, 18*(3), 279–301.

McDonnell, M. (2001). E-collaboration: Transforming your supply chain into a dynamic trading community. *Supply Chain Practice, 3*(2), 80–89.

Melville, N., Kraemer, K., & Gurbaxani, V. (2004). Information technology and organizational performance: An integrative model of it business value. *MIS Quarterly, 28*(2), 283–322.

Menon, A., Sundar, G., Bharadwaj, P. T. A., & Edison, S. W. (1999). Antecedents and consequences of marketing strategy making: A model and a test. *Journal of Marketing, 63*(2), 18–40.

Mentzer, J. T., DeWitt, W., Keebler, J. S., Min, S., Nix, N. W., Smith, C. D., et al. (2001). Defining supply chain management. *Journal of Business Logistics, 22*(2), 1–25.

Min, S., Roath, A., Daugherty, P. J., Genchev, S. E., Chen, H., & Arndt, A. D. (2005). Supply chain collaboration: What's happening? *International Journal of Logistics Management, 16*(2), 237–256.

Mohr, J., Fisher, R. J., & Nevin, J. R. (1996). Collaborative communication in interfirm relationships: Moderating effects of integration and control. *Journal of Marketing, 60*(3), 103–115.

Mohr, J., & Nevin, J. R. (1990). Communication strategies in marketing channels: A theoretical perspective. *Journal of Marketing, 54*(4), 36–51.

Mohr, J. J., & Sohi, R. S. (1995). Communication flows in distribution channels: Impact on assessments of communication quality and satisfaction. *Journal of Retailing, 71*(4), 393–416.

Mohr, J., & Spekman, R. E. (1994). Characteristics of partnership success: partnership attributes, communication behavior, and conflict resolution techniques. *Strategic Management Journal, 15*(2), 135–152.

Monczka, R., Petersen, K., Handfield, R. B., & Ragatz, G. (1998). Success factors in strategic supplier alliances: the buying company perspective. *Decision Sciences, 29*(3), 15–30.

Moorman, C. (1995). Organizational market information processes: Cultural antecedents and new product outcomes. *Journal of Marketing Research, 32*(3), 318–335.

Narasimhan, R., & Das, A. (1999). Manufacturing agility and supply chain management practices. *Production and Inventory Management Journal, 40*(1), 4–10.

Naude, P., & Buttle, F. (2001). Assessing relationship quality. *Industrial Marketing Management, 29*(4), 351–361.

Nyaga, G., Whipple, J., & Lynch, D. (2010). Examining supply chain relationships: Do buyer and supplier perspectives on collaborative relationships differ? *Journal of Operations Management, 28*(2), 101–114.

Olhager, J., & Selldin, E. (2003). Enterprise resource planning survey of Swedish manufacturing firm. *European Journal of Operational Research, 145*(2), 365–373.

Oliva, R., & Watson, N. (2011). Cross-functional alignment in supply chain planning: A case study of sales and operations planning. *Journal of Operations Management, 29*(5), 434–448.

Pahlberg, C. (1997). Cultural differences and problems in HQ-subsidiary relationships in MNCs. In I. Björkman & M. Forsgren (Eds.), *The nature of the international firm* (pp. 451–473). Copenhagen: Handelshöyskolens Forlag.

Peck, H., & Juttner, U. (2000). Strategy and relationships: Defining the interface in supply chain contexts. *International Journal of Logistics Management, 11*(2), 33–44.

Plambeck, E., Lee, H., & Yatsko, P. (2012). Improving environmental performance in your Chinese supply chain. *Sloan Management Review, 53*(2), 43–51.

Poirier, C. C., & Houser, W. F. (1993). *Business partnering for continuous improvement* (pp. 56–201). San Francisco: Berrett-Koehler.

Powell, W. W., Kogut, K. W., & Smith-Doerr, L. (1996). Interorganizational collaboration and the locus of innovation: Networks of learning in biotechnology. *Administrative Science Quarterly, 41*(1), 116–145.

Prahinski, C., & Benton, W. C. (2004). Supplier evaluations: Communication strategies to improve supplier performance. *Journal of Operations Management, 22*(1), 39–62.

Purdy, L., Astad, U., & Safayeni, F. (1994). Perceived effectiveness of the automotive supplier evaluation process. *International Journal of Operations & Production Management, 14*(6), 91–103.

Quinn, F. J. (1999). Cooperation and collaboration: The keys to supply chain success. *Logistic Management & Distribution Report, 38*(2), 35.

Raghunathan, S. (1999). Interorganizational collaborative forecasting and replenishment systems and supply chain implications. *Decision Sciences, 30*(4), 1053–1071.

Ren, Z., Cohen, M., Ho, T., & Terwiesch, C. (2010). Information sharing in a long term supply chain relationship: the role of customer review strategy. *Operations Research, 58*(1), 81–93.

Ring, P. S., & Van de Ven, A. H. (1992). Structuring cooperative relationships between organizations. *Strategic Management Journal, 13*(7), 483–498.

Roberts, E. B., & Berry, C. A. (1985). Entering new businesses: selecting strategies for success. *Sloan Management Review, 26*(3), 3–17.

Rogers, E., & Agarwala-Rogers, R. (1976). *Communication in organizations*. New York: The Free Press.

Rutner, S. M., Gibson, B. J., & Gustin, C. M. (2001). Longitudinal study of supply chain information systems. *Production and Inventory Management Journal, 42*(2), 49–56.

Ryu, K., & Yucesan, E. (2010). A fuzzy newsvendor approach to supply chain coordination. *European Journal of Operational Research, 200*(2), 421–438.

Sako, M. (1992). *Prices, quality and trust: Inter-firm relations in Britain and Japan*. Cambridge: Cambridge University Press.

Sako, M., & Helper, S. (1998). Determinants of trust in supplier relations: Evidence from the automotive industry in Japan and the United States. *Journal of Economic Behavior & Organization, 34*(3), 387–417.

Saxenian, A. (1991). The origins and dynamics of production networks in silicon valley. *Research Policy, 20*.

Saxton, T. (1997). The effects of partner and relationship characteristics on alliance outcomes. *Academy of Management Journal, 40*(2), 443–461.

Scott-Morton, M. S. (Ed.). (1991). *The corporation of the 1990s: Information technology and organizational transformation*. New York: Oxford University Press.

Sheu, C., Yen, H. R., & Chae, D. (2006). Determinants of supplier-retailer collaboration: Evidence from an international study. *International Journal of Operations & Production Management, 26*(1), 24–49.

Simatupang, T. M., & Sridharan, R. (2002). The collaborative supply chain. *International Journal of Logistics Management, 13*(1), 15–30.

Simatupang, T.M., & Sridharan, R. (2004). A benchmarking scheme for supply chain collaboration. *Benchmarking: An International Journal, 11*(1), 9–30.

Simatupang, T. M., & Sridharan, R. (2005a). An Integrative framework for supply chain collaboration. *International Journal of Logistics Management, 16*(2), 257–274.

Simatupang, T. M., & Sridharan, R. (2005b). Supply chain discontent. *Business Process Management Journal, 11*(4), 349–369.

Simatupang, T. M., & Sridharan, R. (2005c). The collaboration index: A measure for supply chain collaboration. *International Journal of Physical Distribution and Logistics Management, 35*(1), 44–62.

Simatupang, T. M., Wright, A. C., & Sridharan, R. (2002). The knowledge of coordination for supply chain integration. *Business Process Management Journal, 8*(3), 289–308.

Simonin, B. L. (1997). The importance of collaborative know-how: An empirical test of the learning organization. *Academy of Management Journal, 40*(2), 1150–1174.

Simpson, J. T., & Mayo, D. T. (1997). Relationship management: A call for fewer influence attempts? *Journal of Business Research, 39*(3), 209–218.

Simpson, P. M., Sigauw, J. A., & White, S. C. (2002). Measuring the performance of suppliers: An analysis of evaluation process. *Journal of supply Chain Management, 38*(1), 29–41.

Skjoett-Larsen, T., Thernoe, C., & Andersen, C. (2003). Supply chain collaboration. *International Journal of Physical Distribution and Logistics Management, 33*(6), 531–549.

Slater, S. F., & Narver, J. C. (1995). Market orientation and the learning organization. *Journal of Marketing, 59*(3), 63–74.

Sobrero, M., & Roberts., E. (2001). The trade-off between efficiency and learning in interorganizational relationships for product development. *Management Sciences, 47*(4), 493–511.

Srinivasan, K., Kekre, S., & Mukhopadhyay, T. (1994). Impact of electronic data interchange technology on JIT shipments. *Management Science, 40*(10), 1291–1304.

Sriram, V., Krapfel, R., & Spekman, R. E. (1992). Antecedents to buyer-seller collaboration: An analysis from the buyer's perspective. *Journal of Business Research, 25*(4), 303–320.

Srivastava, R. K., Shervani, T. A., & Fahey, L. (1998). Market-based assets and shareholder value: A framework for analysis. *Journal of Marketing, 62*(1), 2–18.

Stank, T. P., Daugherty, P. J., & Ellinger, A. E. (1999). Marketing/Logistics integration and firm performance. *International Journal of Logistics Management, 10*(1), 11–23.

Stank, T. P., Keller, S. B., & Daugherty, P. J. (2001). Supply chain collaboration and logistical service performance. *Journal of Business Logistics, 22*(1), 29–48.

Stock, G. N., & Tatikonda, M. H. (2000). A conceptual typology of project-level technology transfer processes. *Journal of Operations Management, 18*(6), 719–737.

Stohl, C., & Redding, W. C. (1987). Messages and message exchange processes. In F. Jablin, et al. (Eds.), *Handbook of organizational communication: An interdisciplinary perspective* (pp. 451–502). Newbury Park: Sage Publications, Inc.

Stuart, F. I. (1997). Supplier alliance success and failure: A longitudinal dyadic perspective. *International Journal of Production and Operations Management, 17*(6), 539–557.

Stuart, F. I., & McCutcheon, D. (1996). Sustaining strategic supplier alliances. *International Journal of Operation and Production Management, 16*(10), 5–22.

Tjosvold, D. (1986a). Dynamics and outcomes of goal interdependencies in organizations. *Journal of Psychology, 120*, 101–112.

Tjosvold, D. (1986b). The dynamics of interdependence in organizations. *Human Relations, 39*(6), 517–540.

Tuten, T. L., & Urban, D. J. (2001). An expanded model of business-to-business partnership foundation and success. *Industrial Marketing Management, 30*(2), 149–164.

Tyndall, G., Gopal, C., Partsch, W., & Kamauff, J. (1998). *Supercharging supply chains: New ways to increase value through global operational excellence.* New York: John Wiley and Sons.

Uzzi, B. (1997). Social structure and competition in interfirm networks: The paradox of embeddedness. *Administrative Science Quarterly, 42*(1), 35–67.

Verdecho, M., Alfaro-Saiz, J., Rodriguez–Rodriguez, R., & Ortiz-Bas, A. (2012). A multi-criteria approach for managing inter-enterprise collaborative relationships. *Omega, 40*(3), 249–263.

Verwaal, E., & Hesselmans, M. (2004). Drivers of supply network governance: An explorative study of the Dutch chemical industry. *European Management Journal, 22*(4), 442–451.

Von Hippel, E. (1988). *The sources of innovation.* New York: Oxford University Press.

Waller, M. A., Dabholkar, P. A., & Gentry, J. J. (2000). Postponement, product customization, and market-oriented supply chain management. *Journal of Business Logistics, 21*(2), 133–160.

Walton, R. E. (1996). A theory of conflict in lateral organizational relationships. In J. R. Lawrence (Ed.), *Operational Research and the Social Science.* London: Tavistock.

Walton, S. V., & Marucheck, A. S. (1997). The relationship between EDI and supplier reliability. *International Journal of Purchasing and Materials Management, 33*(3), 30–35.

Watson, G. (2001). SubRegimes of power and integrated supply chain management. *Journal of Supply Chain Management, 37*(2), 36–41.

Whipple, J. M., & Frankel, R. (2000). Strategic alliance success factors. *Journal of Supply Chain Management, 36*(3), 21–28.

Womack, J. P., Jones, D. T., & Roos, D. (1990). *The machine that changed the world.* New York: Harper Perennial.

Wong, A. (1999). Partnering through cooperative goals in supply chain relationships. *Total Quality Management, 10*(4/5), 786–792.

# Chapter 5
# Collaborative Advantage as Consequences

**Abstract** Supply chain collaboration is rooted in a paradigm of collaborative advantage rather than competitive advantage. Collaborative advantage comes from relational rents that produce common benefits for bilateral rent-seeking behaviors while competitive advantage encourages individual rent-seeking behaviors that maximize a firm's own benefits. The perspective of collaborative advantage enables supply chain partners to view supply chain collaboration as a positive-sum game rather than a zero-sum game where partners strive to appropriate more relational rents for their own competitive advantage. Although collaborative advantage is acknowledged in the literature, there is no operationalization that has been done. In this chapter, we define and operationalize collaborative advantage and firm performance as consequences of supply chain collaboration. Five elements of collaborative advantage (i.e., process efficiency, offering flexibility, business synergy, quality, and innovation) have been investigated. We also develop hypotheses based on the framework proposed in the previous chapters.

As consequences of supply chain collaboration, collaborative advantage and firm performance will be discussed in the following section. In addition, hypotheses will be developed in this chapter.

Collaborative advantage is also called joint competitive advantage (Jap 2001). It refers to strategic benefits gained over competitors in the marketplace through supply chain partnering. Such joint competitive advantage resides not within an individual firm, but across the boundaries of a firm via its relationship with supply chain partners (Dyer 1996; Dyer and Singh 1998; Kanter 1994; Jap 2001; Ferratt et al. 1996; Vangen and Huxham 2003; Foss and Nielsen 2010; Corsten et al. 2011). Ferratt et al. (1996) define collaborative advantage as the benefit gained by a group of firms as the result of their cooperation rather than their competition. They argue that, in healthcare industry, IT enables firms to achieve competitive advantage through collaboration not only with supply chain partners but also with competitors (Pouloudi 1999).

M. Cao and Q. Zhang, *Supply Chain Collaboration*,
DOI: 10.1007/978-1-4471-4591-2_5, © Springer-Verlag London 2013

Collaborative advantage relates to the desired synergistic outcome of collaborative activity that could not have been achieved by any firm acting alone (Vangen and Huxham 2003; Mentzer et al. 2001; Stank et al. 2001; Manthou et al. 2004; Sheu et al. 2006; Foss and Nielsen 2010). Jap (1999) explains that collaboration can enlarge the size of the joint benefits and give each member a share of greater gain that could not be generated by each member on its own. Kanter (1994) argues that supply chain partnering, as the strongest and closest collaboration, is a living system that grows progressively in their possibilities. Collaboration involves creating new values together rather than mere exchange, and it is controlled not by formal systems but by a web of links and infrastructures that augment learning and open new doors for unforeseen opportunities. Thus, collaboration-associated benefits may not be immediately visible; however potential long-term rewards are enticing and strategic (Min et al. 2005).

Hansen and Nohria (2004) argue it is ever harder to sustain competitive advantage based on the economics of scale and scope. Competitive advantage will belong to firms that can encourage and stimulate collaboration to leverage isolated resources. They contend that the value creation from collaboration could be cost savings by way of best practices sharing, enhanced capacity and flexibility for collective actions, better decision making and increased revenue through recourse synergy, and innovation through the integration of ideas. Similarly, Lado et al. (1997) and Luo et al. (2006) suggest that collaboration produces various benefits including cost savings, resource sharing, learning, and innovation.

Synthesizing the above studies, this research conceptualizes collaborative advantage as the following five sub-components: process efficiency, offering flexibility, business synergy, quality, and innovation (Table 5.1). These collaborative advantage and performance are viewed from the perspective of an individual supply chain member. More specifically, the focus concerns the focal firm's overall view of the performance outcomes of supply chain relationships (Duffy and Fearne 2004).

## 5.1  Process Efficiency

Process efficiency refers to the extent to which a firm's collaboration with supply chain partners is cost competitive (Bagchi and Skjoett-Larsen 2005; Simatupang and Sridharan 2005a). The process could be information sharing process, joint logistics process, joint product development process, or joint decision making process. Process efficiency is a measure of success and a determinant factor of the firm's ability to profit (e.g., inventory turnover and operating cost). Supply chain collaboration facilitates the cooperation of participating members along the supply chain to improve performance (Bowersox 1990). The benefits of collaboration include cost reductions and revenue enhancements (Fisher 1997; Lee et al. 1997; Simatupang and Sridharan 2005a).

**Table 5.1** Definition of collaborative advantage and sub-components

| Construct | Definition | Literature |
|---|---|---|
| Collaborative advantage | Strategic benefits gained over competitors in the marketplace through supply chain partnering | Jap (2001), Dyer (1996), Dyer and Singh (1998), Ferratt et al. (1996), Kanter (1994) and Vangen and Huxham (2003) |
| Process efficiency | The extent to which a firm's collaboration with supply chain partners is cost competitive | Bagchi and Skjoett-Larsen (2005), Fisher (1997), Lee et al. (1997) and Simatupang and Sridharan (2005a) |
| Offering flexibility | The extent to which a firm's supply chain linkage supports changes in products or services available for customers | Beamon (1998), Gosain et al. (2004), Holweg et al. (2005), Kiefer and Novack (1999) and Narasimham and Jayaram (1998) |
| Business synergy | The extent to which supply chain partners combine complementary and related resources to achieve spill-over benefits | Ansoff (1988), Itami and Roehl (1987), Larsson and Finkelstein (1999), Lasker et al. (2001), Tanriverdi (2006) and Zhu (2004) |
| Quality | The extent to which a firm with supply chain partners offers reliable and durable products that create higher value for customers | Gray and Harvey (1992), Li (2002) and Dai et al. (2012) |
| Innovation | The extent to which a firm works jointly with its supply chain partners in introducing new processes, products, or services | Clark and Fujimoto (1991), Dyer and Singh (1998), Handfield and Pannesi (1995), Kessler and Chakrobarti (1996), Malhotra et al. (2001), Mowery and Rosenberg (1998), Nishiguchi and Anderson (1995), Rosenblum and Spencer (1996), Sapolsky et al. (1999) and Vesey (1991) |

## 5.2 Offering Flexibility

Offering flexibility refers to the extent to which a firm's supply chain linkage supports changes in products or services available for customers. It is also called customer responsiveness in literature (Beamon 1998; Narasimham and Jayaram 1998; Kiefer and Novack 1999; Holweg et al. 2005). Supply chain partners should be able to change offerings (e.g., features, volume, and speed) in response to environmental changes. Offering flexibility is based on the ability of collaborating firms to quickly change process structures or adapt the information sharing process for modifying the features of a product or service (Gosain et al. 2004). In today's market firms indeed pay attention to customers and more firms solicit customer inputs at the design stage resulting in better acceptance of the products and services later (Bagchi and Skjoett-Larsen 2005).

## 5.3 Business Synergy

Business synergy refers to the extent to which supply chain partners combine complementary and related resources to achieve spill-over benefits. Ansoff (1988) suggests that synergy can produce a combined return on resources that is larger than the sum of individual parts (e.g., $2 + 2 = 5$). This joint effect results from the better use of resources in the supply chain, including physical assets (e.g., facilities, computers, and networks) and invisible assets (e.g., knowledge, expertise, and culture) (Itami and Roehl 1987). Tanriverdi (2006) offers two major sources of synergy: super-additive value by complementary resources and sub-additive cost (or economies of scope) by related resources. Collaboration can help partners to maximize their assets utilization (e.g. truckload transportation and transportation capacity sharing) resulting in substantial capital relief (Min et al. 2005).

Lasker et al. (2001) claim that synergies between supply chain partners are more than a mere exchange of resources. By combining the individual firms' resources, skills, and social capital, the collaboration can create something new and valuable together. Supply chain partners can also achieve synergy of common IT infrastructure, common IT management processes, and common IT vendor management processes (Larsson and Finkelstein 1999; Zhu 2004; Tanriverdi 2006). As long as supply chain partners make decisions in the best economic interest of the whole supply chain, not its own portion, the gain or joint outcome will be expanded (Simatupang and Sridharan 2005a).

## 5.4 Quality

Quality refers to the extent to which a firm with supply chain partners offers reliable and durable products that create higher value for customers (Gray and Harvey 1992; Li 2002; Dai et al. 2012). It is expected that firms that can respond fast to customer needs with high quality product and innovative design, and excellent after-sales service allegedly build customer loyalty, increase market share and ultimately gain high profits. Garvin (1988) proposes eight dimensions of quality: performance, features, reliability, conformance, durability, serviceability, aesthetics, and perceived quality, which are comprehensive but measures for each are difficult to establish.

## 5.5 Innovation

Innovation refers to the extent to which a firm works jointly with its supply chain partners in introducing new processes, products, or services. Due to shorter product life cycles, firms need to innovate frequently and in small increments (Clark and Fujimoto 1991; Vesey 1991; Handfield and Pannesi 1995; Kessler and

Chakrobarti 1996). By carefully managing their relationships with suppliers and customers, firms improve their ability to engage in process and product innovation (Zammuto and O'Connor 1992; Hage 1999; Kaufman et al. 2000). Innovation as a highly structured, knowledge-intensive activity embeds in networks that span organizational and geographical boundaries (Nishiguchi and Anderson 1995; Rosenblum and Spencer 1996; Dyer and Singh 1998; Mowery and Rosenberg 1998; Sapolsky et al. 1999; Malhotra et al. 2001). By tapping joint creativity capacities, joint organizational learning, knowledge sharing, joint problem solving between supply chain partners, firms can improve absorptive capacity and thus introduce new products and services fast and frequently.

## 5.6 Firm Performance

Firm performance refers to how well a firm fulfills its market and financial goals compared with the firm's primary competitors (Tan et al. 1998; Yamin et al. 1999; Li 2002; Barua et al. 2004). In this study firm performance is measured by market share, growth of market share, sales growth, profit margin on sales, return on investment (ROI), growth in return on investment, and overall competitive position. These measures have been extensively employed in previous studies because they are primary yardsticks for most stakeholders (Cooper and Kleischmeidt 1994; Loch et al. 1996; Vickery et al. 1999; Stock et al. 2000; Chang and King 2005; Chen and Paulraj 2004; Flynn et al. 2010; Petersen et al. 2005; Narasimhan and Kim 2002). Effectiveness of supply chain collaboration should be reflected on such financial metrics.

## 5.7 Hypotheses Development

Based on multiple theories, the framework (Fig. 2.1) that relates constructs of IT resources, IOS appropriation, collaborative culture, trust, SC collaboration, collaborative advantage, and firm performance has been developed to conjecture probable truth. In the following sections, hypotheses proposed in the framework will be discussed.

Researchers long argued that IT resources directly lead to better organizational performance (Rockart et al. 1996; Ross et al. 1996; Bharadwaj 2000; Santhanam and Hartono 2003). However, IT resources are not directly converted into measurable outcomes for the organization (McKeen et al. 2005). IT resources support different levels of IOS use by providing flexible IT infrastructure, technical IT skills, and managerial IT knowledge. It is the patterns of IT use (i.e., IOS appropriation), which facilitate collaboration among supply chain partners, that enables conversion effectiveness and actually transforms IT assets into economic and social values (Weill 1992; Markus and Soh 1993; Piccoli and Ives 2005). Therefore, this study develops the following hypothesis:

*Hypothesis* 1: *IT Resources has a significant positive effect on IT appropriation.*

There are three types of IOS appropriation that are critical for supply chain collaboration. First, IOS use for communication enables frequent and bidirectional contact and message flow. IOS technologies such as email, fax, instant messaging, electronic bulletin board, voice mail, Computer Supported Collaborative Working (CSCW) make communication between partners easy, fast, and rich, therefore, partners can work together anytime, anywhere, share real-time information and make better decisions (Bafoutsou and Mentzas 2002). Better communication also provides a more effective platform for supply chain partners to engage in coordination, participation, and problem solving activities (Sheu et al. 2006). Kalafatis et al. (2000) indicates there is a positive relationship between better communication and supplier-retailer collaboration.

Second, IOS use for intelligence (such as shared data repository, data warehouse, data mining, intelligent agents) facilitate joint learning, decision making, and joint knowledge creation (Milton et al. 1999; O'Leary 2003; Tsui 2003). Third, IOS use for integration (e.g., EDI) provides visibility and transparency to supply chain partners and thus it allows intensive information sharing, joint planning, and better execution by electronically coupling business processes between partners (Bensaou and Venkatraman 1995; Barua et al. 2004). Suppliers tend to maintain closer relationships with the customer when they make a higher degree of transaction-specific investments (Son et al. 2005). The majority of research on the association between IT and collaboration proposes a positive link between EDI and buyer–supplier relations (Emmelhainz 1988; Larson and Kulchitsky 2000).

Successful supply chain collaboration depends largely on partners' implementation of the IOS technology (Son et al. 2005). Information technologies have increased the propensity for collaboration by allowing interfirm computer-integrated manufacturing (Adler 1988; Chesborough and Teece 1996; Argyres 1999; Kaufman et al. 2000). Bensaou (1997) found that cooperation between automakers and their suppliers is positively associated with IT use between the trading partners in the Japanese automobile industry. Malone et al. (1987) contend that the electronic integration between firms can reduce the costs of coordinating economic transactions and production, and thus facilitate collaboration. Thus this study hypothesizes:

*Hypothesis* 2: *IT appropriation has a significant positive effect on supply chain collaboration.*

Firms with collaborative culture are more likely to encourage a long-term relationship with supply chain partners by using IOS to integrate business processes and reduce uncertainty. Collectivists will focus on collective goals, promote frequent communications with available technonologies, and even use data mining and data warehousing tools to jointly explore new useful information and knowledge with their supply chain partners (Kumar et al. 1998). Firms with low power distance are more likely to involve their supply chain partners to pull in technologies for joint knowledge discovery and joint decision making (Bates et al. 1995; Hofstede 1980). Thus, the following hypothesis is derived:

**Hypothesis 3$_a$:** *Collaborative culture has a significant positive effect on IOS appropriation.*

Firms with collaborative culture will encourage a long-term relationship with supply chain partners through social norms and trust rather than legal contracts and rigid rules (Walls 1993; Kumar et al. 1998). Collectivists will focus on collective goals rather than unilateral objectives and thus more likely to form cooperative partnerships, encourage frequent communication and intensive information sharing, and solve problems jointly (Wagner 1995). Firms with long-term orientation will be willing to make effort in collaborating by establishing relationship-specific investment (Sheu et al. 2006). Firms with high uncertainty avoidance will be more likely to collaborate with supply chain partners to reduce risk and uncertainty and share cost together.

Power conditions within the supply chain can be either symmetrical (power balanced) or asymmetrical (power imbalance) (Dwyer and Walker 1981). Communication under symmetrical power will have higher frequency and more bidirectional flows, which reduce uncertainty (Mohr and Nevin 1990; Stohl and Redding 1987). Moreover, because the supply chain partners have equal footing in the relationship, they will try to stay abreast of each other's actions (e.g., implementing programs and policies) by frequent communications (Mohr and Nevin 1990). Firms with low power distance are more likely to take on equality, joint decision making, and benefits sharing (Bates et al. 1995; Hofstede 1980; Wuyts and Geyskens 2005; Bagchi and Skjoett-Larsen 2005). Following hypothesis is thus derived from the discussions:

**Hypothesis 3$_b$:** *Collaborative culture has a significant positive effect on supply chain collaboration.*

Trust is an important prerequisite for effective IOS use. If supply chain partners trust each other, they will use technologies and share information openly, communicate easily and frequently, and even jointly explore new knowledge using confidential data and information (Jap 2001; Lejeune and Yakova 2005; Koenig and van Wijk 1994; Kumar et al. 1998).

The lack of trust between top managements of supply chain partners could be a serious problem for interorganizational systems use. If supply chain partners do not trust each other, they will hold back information, and business process will never be integrated even the best technologies and systems are adopted and implemented in place. Thus, this study hypothesizes:

**Hypothesis 4$_a$:** *Trust has a significant positive effect on IOS appropriation.*

In the interorganizational literature, trust is frequently highlighted as key variables that contribute to relationship success (Duffy and Fearne 2004). High level of trust reduces the perceived risk associated with the occasional opportunistic behaviors of partners. Suppliers' perception of the customer's trustworthiness would lead them to establish more cooperative relationships with the customer (Son et al. 2005). Conversely, the lack of trust between the companies' management never develops a long-term orientation and discourages information sharing and IT applications (Sheu et al. 2006).

Trust is a governance mechanism for coordinating interorganizational exchange by implicit social contract, not formal rules (Morgan and Hunt 1994; Jap 2001; Lejeune and Yakova 2005). It diminishes uncertainty in interorganizational exchange through self control (Koenig and van Wijk 1994; Kumar et al. 1998). Moreover, in mutually supportive and trusting climates, communication has higher frequency, more directional flows, and more informal modes (Mohr and Nevin 1990; Blair et al. 1985; Fulk and Mani 1996; Guetzkow 1965; Roberts and O'Reilly 1974). Better communication reduces conflicts and enhances supply chain collaboration. Following hypothesis is thus derived from the discussions:

*Hypothesis 4$_b$: Trust has a significant positive effect on supply chain collaboration*

Previous studies suggested that collaboration (e.g., alliance) benefits include cost reduction, risk sharing, access to financial capital, complementary assets, improved capacity for rapid learning, and knowledge transfer (Eisenhardt and Schoonhoven 1996; Kogut 1988; Powell et al. 1996; Singh and Mitchell 1996; Park et al. 2004). Spekman (1988) holds that buyers are forging closer, more collaborative relationships with a smaller number of vendors to gain greater competitive advantage. Simatupang and Sridharan (2005a) introduce a collaboration index to measure the level of collaborative practices and find that the collaboration index is positively associated with operational performance.

Previous researches also support the finding that information sharing (Frankel et al. 2002; Whipple et al. 2002), joint decision-making (Bowersox 1990; Ramdas and Spekman 2000), and incentive alignment (Narus and Anderson 1996; Corbett et al. 1999) facilitate the process efficiency. Higher levels of collaboration result in operational efficiency in supply chain systems in terms of inventory levels and levels of satisfaction (Sheu et al. 2006).

Supply chain collaboration enables the chain members to create responsiveness to react to demand changes. Close collaboration enables the supply chain partners to improve their ability to fulfill customer needs by flexible offerings (Barratt and Oliveira 2001; Simatupang and Sridharan 2004, 2005b). Decision synchronization and incentive alignment significantly influence responsiveness performance (Fisher 1997; Narus and Anderson 1996; Simatupang and Sridharan 2005b).

Supply chain collaboration promotes a firm's capability to profit quickly from market opportunities (Uzzi 1997). For example, joint problem solving increases the speed-to-market by resolving problems faster. On the basis of the knowledge-based view of the firm, competitive advantage results from innovation enabled by different knowledge stores and market expertise (Grant 1996; Nonaka 1994; Luo et al. 2006). Collaboration between supply chain partners can be sources of new product ideas (Jackson 1985; Weitz et al. 1992; Kalwani and Narayandas 1995).

Shared resources between supply chain partners could be related sources, which reduces sub-additive cost, or complementary resources, which bring super-additive value (Tanriverdi 2006). Both sources of business synergy can bring joint competitive advantage (i.e., collaborative advantage). Supply chain partners are able to expand the total reward due to synergy through collaborative processes (Simatupang and Sridharan 2005a; Jap 1999). Firms such as Procter & Gamble, Hewlett-Packard, IBM, and Dell which work closely with their partners have captured the advantage of collaboration (Barratt and Oliveira 2001; Callioni and Billington 2001; Dell and Fredman 1999). Therefore, this study develops the following hypothesis:

*Hypothesis 5: Supply chain collaboration has a significant positive effect on collaborative advantage.*

Many scholars contend that both customer and supplier firms seek collaborative relationships with each other as a way of improving performance (Duffy and Fearne 2004; Mohr and Spekman 1994; Sheu et al. 2006). Supplier firms can gain great sales and returns from resources invested in developing long-term relationships with their customers (Kalwani and Narayandas 1995). Kalwani and Narayandas (1995) also confirm that suppliers in long-term, closer relationships accomplish more sales growth and profitability compared with those in arm's length bargain relationships with their customers. Stank et al. (2001) suggest that both internal and external collaborations are necessary to ensure performance. Partnerships can improve profitability, reduce purchasing costs, and increase technical cooperation (Ailawadi et al. 1999; Han et al. 1993).

Lee and Whang (2001) report a study performed jointly by Stanford University and Accenture (formerly Andersen Consulting) on 100 manufacturers and 100 retailers in the consumer products and food industry. It reveals that companies that were engaged in higher levels of information sharing reported higher than average profits. In general, researchers suggest that the higher the level of interdependence (i.e., higher level of collaboration) in a relationship, the better firm performance (Duffy and Fearne 2004; Mohr and Spekman 1994; Gattorna and Walters 1996). Thus this study hypothesizes:

*Hypothesis 6: Supply chain collaboration has a significant positive effect on firm performance*

The necessary condition for supply chain collaboration is that the supply chain partners are able to expand the total gain (e.g., higher revenues) due to synergy (Simatupang and Sridharan 2005a). The supply chain partners will gain financial benefits by increasing responsiveness, especially for innovative products (Fisher 1997). The literature also supports the ability of partnerships to achieve cost savings and reduce duplication of efforts by the firms involved (Herbing and O'Hara 1994; Whipple et al. 1996; Zinn and Parasuraman 1997; Lambert et al. 2004). In particular, cooperation among competitors can foster greater knowledge seeking and result in synergetic rents (Lado et al. 1997).

In the short and medium term, firms will observe improvements in operations (e.g., productivity) as the major payback. In the long run, supply chain collaboration will enable faster product development that will be transformed into competitive advantage and increased profits and market share (Stuart and McCutcheon 1996). Thus this study hypothesizes:

> *Hypothesis 7: Collaborative advantage has a significant positive effect on firm performance.*

# References

Adler, P. S. (1988). Managing flexible automation. *California Management Review, 30*(3), 34–56.

Ailawadi, K. L., Farris, P. W., & Parry, M. E. (1999). Market share and ROI: Observing the effect of unobserved variables. *International Journal of Research in Marketing, 16*(1), 17–33.

Ansoff, H. I. (1988). *The new corporate strategy*. New York, NY: Wiley.

Argyres, N. S. (1999). The impact of information technology on coordination: Evidence from the B-2 stealth bomber. *Organization Science, 10*(2), 162–180.

Bafoutsou, G., & Mentzas, G. (2002). Review and functional classification of collaborative systems. *International Journal of Information Management, 22*(4), 281–306.

Bagchi, P. K., & Skjoett-Larsen, T. (2005). Supply chain integration: A survey. *The International Journal of Logistics Management, 16*(2), 275–294.

Barratt, M., & Oliveira, A. (2001). Exploring the experiences of collaborative planning initiatives. *International Journal of Physical Distribution and Logistics Management, 31*(4), 66–89.

Barua, A., Konana, P., Whinston, A. B., & Yin, F. (2004). An empirical investigation of net-enabled business value. *MIS Quarterly, 28*(4), 585–620.

Bates, K. A., Amundson, S. D., Schroeder, R. G., & Morris, W. T. (1995). The crucial interrelationship between manufacturing strategy and organizational culture. *Management Science, 41*(10), 1565–1580.

Beamon, B. M. (1998). Supply chain design and analysis: Models and methods. *International Journal of Production Economics, 55*(3), 281–294.

Bensaou, M. (1997). Interorganizational cooperation: The role of information technology, an empirical comparison of U.S. and Japanese supplier relations. *Information Systems Research, 8*(2), 107–124.

Bensaou, M., & Venkatraman, N. (1995). Configurations of interorganizational relationships: A comparison between U.S. and Japanese automakers. *Management Science, 41*(9), 1471–1492.

Bharadwaj, A. S. (2000). A resource based perspective on information technology capability and firm performance: An empirical investigation. *MIS Quarterly, 24*(1), 169–196.

Blair, R., Roberts, K., & McKechnie, P. (1985). Vertical and network communication in organizations: The present and the future. In R. McPhee & P. Tompkins (Eds.), *Organizational communication: Traditional themes and new directions*. Beverly Hills, CA: Sage Publications, Inc.

Bowersox, D. J. (1990). The strategic benefits of logistics alliances. *Harvard Business Review, 68*(4), 36–43.

Callioni, G., & Billington, C. (2001). Effective collaboration: Hewlett-Packard takes supply chain management to another level. *OR/MS Today, 28*(5), 34–39.

Chang, C. J., & King, W. R. (2005). Measuring the performance of information systems: A functional scorecard. *Journal of Management Information Systems, 22*(1), 85–115.

Chen, I. J., & Paulraj, A. (2004). Towards a theory of supply chain management: The constructs and measurements. *Journal of Operations Management, 22,* 119–150.

Chesborough, H., & Teece, D. (1996). When is virtual virtuous? Organizing for innovation. *Harvard Business Review, 74*(1), 65–73.

Clark, K. B., & Fujimoto, T. (1991). *Product development performance: Strategy, organization, and management in the World auto industry.* Boston, MA: Harvard Business School Press.

Cooper, R. G., & Kleinschmidt, E. J. (1994). Determinants of timeliness in product development. *Journal of Product Innovation Management, 11*(5), 381–396.

Corbett, C. J., Blackburn, J. D., & Van Wassenhove, L. N. (1999). Partnerships to improve supply chains. *Sloan Management Review, 40*(4), 71–82.

Corsten, D., Gruen, T. W., & Peyinghaus, M. (2011). The effects of supplier-to-buyer identification on operational performance: An empirical investigation of inter-organizational identification in automotive relationships. *Journal of Operations Management, 29*(6), 549–560.

Dai, Y., Zhou, S., & Xu, Y. (2012). Competitive and collaborative quality and warranty management in supply chains. *Production and Operations Management, 21*(1), 129–144.

Dell, M., & Fredman, C. (1999). *Direct from Dell: Strategies that revolutionized an industry.* London: Harper Collins Business.

Duffy, R., & Fearne, A. (2004). The impact of supply chain partnerships on supplier performance. *International Journal of Logistics Management, 15*(1), 57–71.

Dwyer, F. R., & Orville, C. W. (1981). Bargaining in an asymmetrical power structure. *Journal of Marketing, 45*(Winter), 104–115.

Dyer, J. H., & Singh, H. (1998). The relational view: Cooperative strategy and sources of interorganizational competitive advantage. *Academy of Management Review, 23*(4), 660–679.

Dyer, J. H. (1996). Does governance matter? Keiretsu alliances and asset specificity as sources of Japanese competitive advantage. *Organization Science, 7*(6), 649–666.

Eisenhardt, K., & Schoonhoven, C. B. (1996). Resource-based view of strategic alliance formation in entrepreneurial firms: Strategic needs and social opportunities for cooperation. *Organization Science, 7*(2), 136–150.

Emmelhainz, M. A. (1988). Strategic issues of EDI implementation. *Journal of Business Logistics, 9*(2), 55–70.

Ferratt, T. W., Lederer, A. L., Hall, J. M., & Krella, J. M. (1996). Swords and plowshares: Information technology for collaborative advantage. *Information and Management, 30*(3), 131–142.

Fisher, M. L. (1997). What is the right supply chain for your product? *Harvard Business Review, 75*(2), 105–116.

Flynn, B. B., Huo, B., & Zhao, X. (2010). The impact of supply chain integration on performance: A contingency and configuration approach. *Journal of Operations Management, 28*(1), 58–71.

Foss, N., & Nielsen, B. (2010). *Researching collaborative advantage: Some conceptual and multi-level issues.* SMG Working Paper No. 6/2010, Copenhagen Business School, ISBN: 978-87-91815-59-1.

Frankel, R., Goldsby, T. J., & Whipple, J. M. (2002). Grocery industry collaboration in the wake of ECR. *International Journal of Logistics Management, 13*(1), 57–71.

Fulk, J. & Mani, S. (1996). Distortion of communication in hierarchical relationships. In I. L. McLaughlin (Ed.), *Communication yearbook* (Vol. 9, pp. 483–510). Beverly Hills, CA: Sage Publications, Inc.

Garvin, D. (1988). Managing quality. *McKinsey Quarterly, 3,* 61–70.

Gattorna, J. L., & David, W. W. (1996). *Managing the supply chain: A strategic perspective.* Basingstoke: Macmillan Business.

Gosain, S., Malhotra, A., & El Sawy, O. A. (2004). Coordinating for flexibility in e-business supply chains. *Journal of Management Information Systems, 21*(3), 7–45.

Grant, R. M. (1996). Toward a knowledge-based theory of the firm. *Strategic Management Journal, 17*(Special Issue), 109–122.

Gray, J., & Harvey, T. (1992). *Quality value banking: Effective management systems that increase earnings, lower costs, and provide competitive customer service*. New York: Wiley.

Guetzkow, H. (1965). Communications in organizations. In J. March (Ed.), *Handbook of organizations* (pp. 534–573). Chicago: Rand McNally.

Hage, J. T. (1999). Organizational innovation and organizational change. *Annual Reviews of Sociology, 25*, 597–622.

Han, S., Wilson, D. T., & Dant, S. P. (1993). Buyer supplier relationships today. *Industrial Marketing Management, 22*(4), 331–338.

Handfield, R. B., & Pannesi, R. T. (1995). Antecedents of leadtime competitiveness in make-to-order manufacturing firms. *International Journal of Production Research, 33*(2), 511–537.

Hansen, M. T., & Nohria, N. (2004). How to build collaborative advantage. *MIT Sloan Management Review, 46*(1), 22–30.

Herbing, P. A., & O'Hara, B. S. (1994). The future of original equipment manufacturers: A matter of relationships. *Journal of Business and Industrial Marketing, 9*(3), 38–43.

Hofstede, G. (1980). *Culture's consequences: International differences in work-related values*. London: Sage Publications.

Holweg, M., Disney, S., Holmström, J., & Småros, J. (2005). Supply chain collaboration: Making sense of the strategy continuum. *European Management Journal, 23*(2), 170–181.

Itami, H., & Roehl, T. (1987). *Mobilizing invisible assets*. Cambridge, MA: Harvard University Press.

Jackson, B. B. (1985). *Winning and keeping industrial customers*. Lexington, KY: Lexington Books.

Jap, S. D. (2001). Perspectives on joint competitive advantages in buyer-supplier relationships. *International Journal of Research in Marketing, 18*(1/2), 19–35.

Jap, S. D. (1999). Pie-expansion efforts: Collaboration processes in buyer-supplier relationships. *Journal of Marketing Research, 36*(4), 461–476.

Kalafatis, S. P., Tsogas, M. H., & Blankson, C. (2000). Positioning strategies in business markets. *Journal of Business and Industrial Marketing, 15*(6/7), 416–437.

Kalwani, M. U., & Narayandas, N. (1995). Long-term manufacturer–supplier relationships: Do they pay? *Journal of Marketing, 59*(1), 1–15.

Kanter, R. (1994). Collaborative advantage: The art of alliances. *Harvard Business Review, 72*(4), 96–108.

Kaufman, A., Wood, C. H., & Theyel, G. (2000). Collaboration and technology linkages: A strategic supplier typology. *Strategic Management Journal, 21*(6), 649–663.

Kessler, E., & Chakrabarti, A. (1996). Innovation speed: A conceptual model of context, antecedents, and outcomes. *The Academy of Management Review, 21*(4), 1143–1191.

Kiefer, A. W., & Novack, R. A. (1999). An empirical analysis of warehouse measurement systems in the context of supply chain implementation. *Transportation Journal, 38*(3), 18–27.

Koenig, C., & van Wijk, G. (1994). Interorganizational collaboration: Beyond contracts. In *Workshop on schools of thought in strategic management. Beyond fragmentation?* ERASM/ Erasmus University, Internal Report, December 14-15.

Kogut, B. (1988). Joint ventures: Theoretical and empirical perspectives. *Strategic Management Journal, 9*(4), 319–332.

Kumar, K., Van Dissel, H. G., & Bielli, P. (1998). The merchant of Prato revisited: Toward a third rationality of information systems. *MIS Quarterly, 22*(2), 199–226.

Lado, A., Boyd, N. G., & Hanlon, S. C. (1997). Competition cooperation and the search for economic rents: A syncretic model. *Academy of Management Review, 22*(1), 110–141.

Lambert, D. M., Emmelhainz, M. A., & Gardner, J. T. (2004). Supply chain partnerships: Model validation and implementation. *Journal of Business Logistics, 25*(2), 21–42.

Larson, P. D., & Kulchitsky, J. D. (2000). The use and impact of communication media in purchasing and supply management. *Journal of Supply Chain Management, 5*(2), 29–37.

Larsson, R., & Finkelstein, S. (1999). Integrating strategic, organizational, and human resource perspectives on mergers and acquisitions: A case survey of synergy realization. *Organization Science, 10*(1), 1–26.

Lasker, R. D., Weiss, E. S., & Miller, R. (2001). Partnership synergy: A practical framework for studying and strengthening the collaborative advantage. *The Milbank Quarterly, 79*(2), 179–205.

Lee, H. L., & Whang, S. (2001). E-Business and supply chain integration. *Stanford Global Supply Chain Management Forum*, SGSCMF-W2-2001.

Lee, H. L., Padmanabhan, V., & Whang, S. (1997). The bullwhip effect in supply chain. *Sloan Management Review, 38*(3), 93–102.

Lejeune, N., & Yakova, N. (2005). On characterizing the 4 C's in supply china management. *Journal of Operations Management, 23*(1), 81–100.

Li, S. (2002). *An integrated model for supply chain management practice, performance, and competitive advantage.* Unpublished Dissertation, University of Toledo.

Loch, C., Stein, L., & Terwisch, C. (1996). Measuring development performance in electronics industry. *The Journal of Product Innovation Management, 13*(1), 3–20.

Luo, X., Slotegraaf, R. J., & Pan, X. (2006). Cross-functional coopetition: The simultaneous role of cooperation and competition within firms. *Journal of Marketing, 70*(2), 67–80.

Malhotra, A., Majchrzak, A., Carman, R., & Lott, V. (2001). Radical innovation without collocation: A case study and Boeing-Rocketdyne. *MIS Quarterly, 25*(2), 229–249.

Malone, T. W., Yates, J., & Benjamin, R. I. (1987). Electronic markets and electronic hierarchies. *Communications of the ACM, 30*(6), 484–497.

Manthou, V., Vlachopoulou, M., & Folinas, D. (2004). Virtual e-Chain (VeC) model for supply chain collaboration. *International Journal of Production Economics, 87*(3), 241–250.

Markkus, M. L., & Soh, C. (1993). Banking on information technology: Converting IT spending into firm performance. In R. Banker, R. Kauffman, & M. A. Mahmood (Eds.), *Strategic information technology management: Perspectives on organizational growth and competitive advantage* (pp. 375–403). Harrisburg, PA: Idea Group Publishing.

McKeen, J. D., Smith, H. A., & Singh, S. (2005). Developments in practice: A framework for enhancing it capabilities. *Communications of the Association for Information Systems, 15*, 661–673.

Mentzer, J. T., DeWitt, W., Keebler, J. S., Min, S., Nix, N. W., Smith, C. D., et al. (2001). Defining supply chain management. *Journal of Business Logistics, 22*(2), 1–25.

Milton, N., Shadbolt, N., Cottam, H., & Hammersley, M. (1999). Towards a knowledge technology for knowledge management. *International Journal of Human-Computer Studies, 51*(3), 615–641.

Min, S., Roath, A., Daugherty, P. J., Genchev, S. E., Chen, H., & Arndt, A. D. (2005). Supply chain collaboration: What's happening? *International Journal of Logistics Management, 16*(2), 237–256.

Mohr, J., & Nevin, J. R. (1990). Communication strategies in marketing channels: A theoretical perspective. *Journal of Marketing, 54*(4), 36–51.

Mohr, J., & Spekman, R. E. (1994). Characteristics of partnership success: Partnership attributes, communication behavior, and conflict resolution techniques. *Strategic Management Journal, 15*(2), 135–152.

Morgan, R. M., & Hunt, S. D. (1994). The commitment-trust theory of relationship marketing. *Journal of Marketing, 58*(3), 20–38.

Mowery, D., & Rosenberg, N. (1998). *Paths of innovation: Technological change in 20th century America.* Cambridge, UK: Cambridge University Press.

Narasimhan, R., & Jayaram, J. (1998). Causal linkage in supply chain management: An exploratory study of north american manufacturing firms. *Decision Science, 29*(3), 579–605.

Narasimhan, R., & Kim, S. W. (2002). Effect of supply chain integration on the relationship between diversification and performance: Evidence from Japanese and Korean firms. *Journal of Operations Management, 20*(3), 303–323.

Narus, J. A., & Anderson, J. C. (1996). Rethinking distribution: Adaptive channels. *Harvard Business Review, 74*(4), 112–120.

Nishiguchi, T., & Anderson, E. (1995). Supplier and buyer networks. In E. Bowman & B. Kogut (Eds.), *Redesigning the firm* (pp. 65–86). New York: Oxford University Press.

Nonaka, I. (1994). A dynamic theory of organizational knowledge creation. *Organization Science, 5*(1), 14–37.

O'Leary, D. E. (2003). Technologies for knowledge assimilation. In C. W. Holsapple (Ed.), *Handbook on knowledge management* (Vol. 2, pp. 29–46). New York, NY: Springer.

Park, N. K., Mezias, J. M., & Song, J. (2004). A resource-based view of strategic alliances and firm value in the electronic marketplace. *Journal of Management, 30*(1), 7–27.

Petersen, K., Handfield, R., & Ragatz, G. (2005). Supplier integration into new product development: Coordinating product, process, and supply chain design. *Journal of Operations Management, 23*(3/4), 371–388.

Piccoli, G., & Ives, B. (2005). IT-dependent strategic initiatives and sustained competitive advantage: A review and synthesis of the literature. *MIS Quarterly, 29*(4), 747–776.

Pouloudi, A. (1999). Information technology for collaborative advantage in healthcare revisited. *Information and Management, 35*(6), 345–356.

Powell, W. W., Kogut, K. W., & Smith-Doerr, L. (1996). Interorganizational collaboration and the locus of innovation: Networks of learning in biotechnology. *Administrative Science Quarterly, 41*(1), 116–145.

Ramdas, K., & Spekman, R. E. (2000). Chain or shackles: Understanding what drives supply-chain performance. *Interfaces, 30*(4), 3–21.

Roberts, K. H., & O'Reilly, C. A. (1974). Measuring organizational communication. *Journal of Applied Psychology, 59*(3), 321–326.

Rockart, J. F., Earl, M., & Ross, J. W. (1996). Eight imperatives for the new IT organization. *Sloan Management Review, 38*(1), 43–55.

Rosenblum R.S., & Spencer, W.J. (1996). *Engines of innovation: U.S. industrial research at an end of an era*. Boston, MA: Harvard Business School Press.

Ross, J. W., Beath, C. M., & Goodhue, D. L. (1996). Develop long-term competitiveness through IT assets. *Sloan Management Review, 38*(1), 31–42.

Santhanam, R., & Hartono, E. (2003). Issues in linking information technology capability to firm performance. *MIS Quarterly, 27*(1), 125–153.

Sapolsky, H., Gholz, E., & Kaufman, A. (1999). Security lessons from the cold war. *Foreign Affairs, 78*(4), 77–89.

Sheu, C., Yen, H. R., & Chae, D. (2006). Determinants of supplier-retailer collaboration: Evidence from an international study. *International Journal of Operations and Production Management, 26*(1), 24–49.

Simatupang, T.M., & Sridharan, R. (2004). A benchmarking scheme for supply chain collaboration. *Benchmarking: An International Journal, 11*(1), 9–30.

Simatupang, T. M., & Sridharan, R. (2005a). An Integrative framework for supply chain collaboration. *International Journal of Logistics Management, 16*(2), 257–274.

Simatupang, T. M., & Sridharan, R. (2005b). The collaboration index: A measure for supply chain collaboration. *International Journal of Physical Distribution and Logistics Management, 35*(1), 44–62.

Singh, K., & Mitchell, W. (1996). Precarious collaboration: Business survival after partners shut down or form new partnerships. *Strategic Management Journal, 17*(7), 99–115.

Son, J., Narasimhan, S., & Riggins, F. J. (2005). Effects of relational factors and channel climate on EDI Usage in the customer-supplier relationship. *Journal of Management information Systems, 22*(1), 321–353.

Spekman, R. E. (1988). Strategic supplier selection: Understanding long-term buyer relationships. *Business Horizons, 31*(4), 75–81.

Stank, T. P., Keller, S. B., & Daugherty, P. J. (2001). Supply chain collaboration and logistical service performance. *Journal of Business Logistics, 22*(1), 29–48.

Stock, G. N., Greis, N. P., & Kasarda, J. D. (2000). Enterprise logistics and supply chain structure: The role of fit. *Journal of Operations Management, 18*(5), 531–547.

Stohl, C., & Redding, W. C. (1987). Messages and message exchange processes. In F. Jablin, et al. (Eds.), *Handbook of organizational communication: An interdisciplinary perspective* (pp. 451–502). Newbury Park, CA: Sage Publications, Inc.

Stuart, F. I., & McCutcheon, D. (1996). Sustaining strategic supplier alliances. *International Journal of Operation and Production Management, 16*(10), 5–22.

Tan, K. C., Kannan, V. R., & Handfield, R. B. (1998). Supply chain management: Supplier performance and firm performance. *International Journal of Purchasing and Materials Management, 34*(3), 2–9.

Tanriverdi, H. (2006). Performance effects of information technology synergies in multibusiness firms. *MIS Quarterly, 30*(1), 57–77.

Tsui, E. (2003). Tracking the role and evolution of commercial knowledge management software. In C. W. Holsapple (Ed.), *Handbook on knowledge management* (Vol. 2, pp. 5–27). New York, NY: Springer.

Uzzi, B. (1997). Social structure and competition in interfirm networks: The paradox of embeddedness. *Administrative Science Quarterly, 42*(1), 35–67.

Vangen, S., & Huxham, C. (2003). Enacting leadership for collaborative advantage: Dilemmas of ideology and pragmatism in the activities of partnership managers. *British Journal of Management, 14*(Supplement 1), S61–S76.

Vesey, J. T. (1991). The new competitors: They think in terms of speed-to-market. *Academy of Management Executive, 5*(2), 23–33.

Vickery, S., Calantone, R., & Droge, C. (1999). Supply chain flexibility: An empirical study. *The Journal of Supply Chain Management, 35*(3), 16–24.

Wagner, J. A, I. I. I. (1995). Studies of individualism-collectivism: Effects on cooperation. *Academy of Management Journal, 38*(1), 152–172.

Walls, J. (1993). Global networking for local development: Task focus and relationship focus in cross-cultural communication. In L. M. Harasim (Ed.), *Global networks: Computers and international communication*. Cambridge, MA: The MIT Press.

Weill, P. (1992). The relationship between investment in information technology and firm performance: A study of the value manufacturing sector. *Information Systems Research, 3*(4), 307–331.

Weitz, B. A., Castleberry, S. B., & Tanner, J. F. (1992). *Selling—building partnerships*. Boston: Richard D. Irwin, Inc.

Whipple, J. M., Frankel, R., & Frayer, D. J. (1996). Logistical alliance formation motives: Similarities and differences within the channel. *Journal of Marketing Theory and Practice, 4*(2), 26–36.

Whipple, J.M., Frankel, R., & Daugherty, P.J. (2002). Information support for alliances: Performance implications. *Journal of Business Logistics, 23*(2), 67–82.

Wuyts, S., & Geyskens, I. (2005). The formation of buyer–supplier relationships: Detailed contract drafting and close partner selection. *Journal of Marketing, 69*(4), 103–117.

Yamin, S., Gunasekruan, A., & Mavondo, F. T. (1999). Relationship between generic strategy, competitive advantage and firm performance: An empirical analysis. *Technovation, 19*(8), 507–518.

Zammuto, R., & O'Connor, E. (1992). Gaining advanced manufacturing technologies benefits: The role of organizational design and culture. *Academy Management Review, 17*(4), 701–728.

Zhu, K. (2004). The complementarity of information technology infrastructure and e-commerce capability: A resource-based assessment of their business value. *Journal of Management Information Systems, 21*(1), 167–202.

Zinn, W., & Parasuraman, A. (1997). Scope and intensity of logistics-based strategic alliances: A conceptual classification and managerial implications. *Industrial Marketing Management, 26*(2), 137–147.

# Chapter 6
# Structured Interview and Q-Sort

**Abstract** To achieve content validity, the measurement items are created through rigorous and extensive review of literature to cover the domain of the constructs. To pre-assess the reliability, convergent validity, and discriminant validity of the scales, the common pool of items is reviewed and evaluated by practitioners. Structured interviews are conducted to check the relevance and clarity of each sub-construct's definition and the wording of questionnaire items. Then, interviewees are asked to sort out the questionnaire items into corresponding sub-constructs. In this chapter, we discussed the procedure and results for structured interview and Q-sort. Three different measures are used to evaluate the Q-sort results including inter-judge raw agreement score, item placement ratio, and Cohen's Kappa

To test the structural relationships among the constructs proposed in the previous chapters, reliable and valid instruments must be developed. These instruments measure (1) IT resources, (2) IOS appropriation, (3) collaborative culture, (4) trust, (5) supply chain collaboration, (6) collaborative advantage, and (7) firm performance. The instruments to measure firm performance were adopted from Li (2002).

The development of instruments for the remaining six constructs was carried out in three steps: (1) item generation, (2) structured interview and Q-sort, and (3) large-scale analysis. First, to ensure the content validity of the constructs, an extensive literature review, as discussed in previous chapters, was conducted to define each construct and generate the initial items for measuring the construct. Then, a pilot study was conducted using structured interview and Q-sort method to provide a preliminary assessment of the reliability, convergent validity, and discriminant validity of the scales. The third step was a large-scale survey to validate the instruments (to be discussed in Chap. 7).

M. Cao and Q. Zhang, *Supply Chain Collaboration*,
DOI: 10.1007/978-1-4471-4591-2_6, © Springer-Verlag London 2013

## 6.1 Item Generation

The objective of item generation is to achieve the content validity of constructs by reviewing literature and consulting with academic and industrial experts. The measurement items for a scale should cover the content domain of a construct (Churchill 1979; Moore and Benbasat 1991; Segars and Grover 1998). To generate measurement items for each construct in the study, prior research was extensively reviewed and an initial list of potential items was compiled. The strategy was to use as few required items as possible to reliably measure the construct based on its definition. A five-point Likert scale was used to indicate the extent to which managers agree or disagree with each statement where 1 = strongly disagree, 2 = disagree, 3 = neutral, 4 = agree, and 5 = strongly agree.

To achieve the content validity for IT resources, previous literature was reviewed (e.g., Ross et al. 1996; Weill et al. 1996; Armstrong and Sambamurthy 1999; Bharadwaj 2000; Byrd and Turner 2000; Dehning and Richardson 2002; Melville et al. 2004; Peppard and Ward 2004; Ranganathan et al. 2004; Bhatt and Grover 2005; Piccoli and Ives 2005; Ravichandran and Lertwongsatien 2005; Ray et al. 2005). Based on the definitions presented in Table 3.1, 14 items were developed to measure IT resources as the bundles of IT assets and capabilities that can be used to support IOS use in supply chain collaboration. These initial items were developed with two scales in mind.

Items for IOS appropriation were developed based on a rigorous review of available literature (e.g., Bensaou and Venkatraman 1995; Mehra and Nissen 1998; Milton et al. 1999; Bafoutsou and Mentzas 2002; Christiaanse and Venkatraman 2002; Grover et al. 2002; Mukhopadhyay and Kekre 2002; Salisbury et al. 2002; Barua et al. 2004; Manthou et al. 2004; Subramani 2004; Chi and Holsapple 2005; Malhotra et al. 2005; Saeed et al. 2005; Nissen and Sengupta 2006). Based on the definitions provided in Table 3.2, 15 items were developed to represent the extent of IOS use. Items were expected to measure three groups corresponding to the three sub-dimensions proposed in the previous chapters.

Items for collaborative culture were generated by reviewing the relevant literature (e.g., Hofstede 1980, 1991, 2001; Bates et al. 1995; Dyer 1996; Nooteboom et al. 1997; Kumar et al. 1998; Kanji and Wong 1999; Boddy et al. 2000; Steensma et al. 2000; Angeles and Nath 2001; Tuten and Urban 2001; Narayandas and Rangan 2004; Verwaal and Hesselmans 2004; Min et al. 2005; Son et al. 2005; Wuyts and Geyskens 2005; Sheu et al. 2006). Based on the definitions proposed in Table 3.3 and the reliable and valid measures used in the past research (e.g., Hofstede 1980, 1991; Kanji and Wong 1999; Wuyts and Geyskens 2005), 16 items were developed to measure the four different aspects of collaborative culture.

Items for trust were generated by reviewing relevant literature (e.g., Ring and Van de Ven 1992; Bhattacharya et al. 1998; Zaheer et al. 1998; Das and Teng 1998; Angeles and Nath 2001; Nesheim 2001; Tuten and Urban 2001; Ba and Pavlou 2002; McKnight and Chervany 2002; Pavlou 2002a, b; Scheer et al. 2003; Johnson et al. 2004; Pavlou and Gefen 2004; De Wever et al. 2005). Based on the definitions proposed in Table 3.4 and the reliable and valid measures adapted from the past research (e.g., Angeles and Nath 2001; Pavlou 2002a, b; Scheer et al. 2003; Johnson et al. 2004), ten items were developed to measure the two aspects of trust.

To develop the items to measure supply chain collaboration, prior literature was thoroughly reviewed (Mohr and Nevin 1990; Sriram et al. 1992; Poirier and Houser 1993; Ellram 1995; Burnes and New 1996; Ellram and Edis 1996; Lambert et al. 1996, 1999; Cooper et al. 1997a, b; Macbeth 1998; Kanji and Wong 1999; Kaufman et al. 2000; Angeles and Nath 2001, 2003; Mentzer et al. 2001; McDonnell 2001; Nesheim 2001; Stank et al. 2001; Bafoutsou and Mentzas 2002; Johnson and Whang 2002; Bowersox et al. 2003; Golicic et al. 2003; Grieger 2003; Hardy et al. 2003; Johnson and Sohi 2003; Harland et al. 2004; Manthou et al. 2004; Marquez et al. 2004; Melville et al. 2004; Prahinski and Benton 2004; Kock and Nosek 2005; Lejeune and Yakova 2005; Malhotra et al. 2005; Simatupang and Sridharan 2005a, b, c; Luo et al. 2006; Sheu et al. 2006). The literature provided a rich pool of items for supply chain collaboration. Out of the extensive literature, 42 items were developed for seven sub-constructs.

Items for collaborative advantage were adapted from previous literature (e.g., Clark and Fujimoto 1991; Kanter 1994; Handfield and Pannesi 1995; Dyer 1996; Ferratt et al. 1996; Fisher 1997; Lee et al. 1997; Dyer and Singh 1998; Jap 2001; Malhotra et al. 2001; Vangen and Huxham 2003; Gosain et al. 2004; Zhu 2004; Bagchi and Skjoett-Larsen 2005; Simatupang and Sridharan 2005a; Tanriverdi 2006). Based on the definitions proposed in Table 5.1, 20 items were developed to measure the five aspects of collaborative advantage.

In summary, there are a total of 23 constructs and 117 items shown in Table 6.1.

## 6.2   Structured Interview and Q-Sort

After the measurement items were created through vigorous and extensive review of literature, the common pool of items were reviewed and evaluated by practitioners from four different manufacturing firms to pre-assess the reliability, convergent validity, and discriminant validity of the scales. First, structured interviews were conducted to check the relevance and clarity of each sub-construct's definition and the wording of questionnaire items. Then, interviewees were asked to sort out the questionnaire items into corresponding sub-constructs. The objective of Q-sort was to pre-assess the convergent and discriminant validity of the scales. Based on the feedback from the experts, redundant and ambiguous items were eliminated or modified. New items were added when necessary.

The basic procedure ran as follows: First, the interviewees were shown the conceptual model and the definition of each construct and sub-construct and were asked whether the model and constructs made sense to them. Then, the interviewees acted as judges and sorted the pool of questionnaire items into separate sub-constructs. The items were divided into two pools because it would be difficult for a judge to sort too many items in one pool. The first pool consisted of items measuring the eleven

**Table 6.1** Constructs, sub-constructs, and number of items

| Constructs | Sub-constructs | # of items |
|---|---|---|
| IT resources | IT infrastructure flexibility | 6 |
| | IT expertise | 8 |
| IOS appropriation | IOS use for integration | 5 |
| | IOS use for communication | 5 |
| | IOS use for intelligence | 5 |
| Collaborative culture | Collectivism | 4 |
| | Long term orientation | 4 |
| | Power symmetry | 4 |
| | Uncertainty avoidance | 4 |
| Trust | Credibility | 5 |
| | Benevolence | 5 |
| Supply chain collaboration | Information sharing | 6 |
| | Goal congruence | 6 |
| | Decision synchronization | 6 |
| | Resource sharing | 6 |
| | Incentive alignment | 6 |
| | Collaborative communication | 6 |
| | Joint Knowledge Creation | 6 |
| Collaborative advantage | Process efficiency | 4 |
| | Offering flexibility | 4 |
| | Business synergy | 4 |
| | Quality | 4 |
| | Innovation | 4 |
| Total | | 117 |

sub-constructs of the constructs: IT resources (2), IOS appropriation (3), collaborative culture (4), and trust (2). The second pool consisted of items measuring the twelve sub-constructs of the constructs: supply chain collaboration (7) and collaborative advantage (5). Each item was printed on a 3 × 5-inch index card. The cards were shuffled into random order for presentation to the judges. Based on their judgment, the judges sorted the cards into separate categories, each category corresponding to a sub-construct. A "Not Applicable" category was included to ensure that the judges did not force any items into a particular category. The judges were allowed to ask any questions related to model, definitions, and procedures to ensure that they understood the procedures correctly. The items were subjected to two sorting rounds by two independent judges per round. The judges were: (1) a material manager of an industrial equipment firm, (2) a plant manager of a leather product firm, (3) a vice president of a transportation equipment firm, and (4) an IT director of an electronic firm.

To assess the reliability of items, three different measures were taken: (1) The inter-judge raw agreement scores are calculated by counting the number of items that both judges agreed to place into certain category, although the category into which items were sorted by both judges might not be the intended one, and dividing it by the total number of items; (2) Item placement ratios are calculated by counting all the items that were correctly sorted into the intended theoretical category by each of the judges, and dividing it by twice the total number of items. It is an indicator of how many items were placed in the intended, or target, categories by the judges; (3) Cohen's Kappa is calculated to measure the level of agreement between the two judges in categorizing the items. It can be interpreted as the proportion of joint judgments in which there is agreement after chance agreement is excluded. A description of the Cohen's Kappa concept and methodology is included in the Appendix B.

In the first round, for the first pool, the inter-judge raw agreement scores average 80 % (Table 6.2), the overall placement ratio of items is 83 % (Table 6.3), and Kappa scores average 0.78 (Table 6.10). Based on the guidelines of Landis and Koch (1977) for interpreting the Kappa coefficient, the value of 0.78 indicates an excellent level of agreement. However, the item placement ratio values for IT infrastructure flexibility, IT expertise, and collectivism were 75, 79, and 75 % respectively, indicating a low degree of construct validity and a need for further improvement. For the second pool, the inter-judge raw scores average 81 % (Table 6.4), the overall placement ratio of items is 82 % (Table 6.5), and Kappa scores average 0.80 (Table 6.10). Based on the guidelines of Landis and Koch (1997), the value of 0.80 indicates an excellent level of agreement. However, there are 6 subcomponents with low item placement ratios, either 67 or 75 %, indicating a low degree of construct validity and a need for further improvement.

In order to improve the Cohen's Kappa measure of agreement, an examination of the off-diagonal entries in the placement matrix (Tables 6.3 and 6.5) was conducted. Any ambiguous items (fitting in more than one category) or too indeterminate items (fitting in no category) were reworded or eliminated. For the first pool, two items were deleted and three items were reworded. For the second pool,

**Table 6.2** Inter-judge raw agreement scores: the first sorting round, the first pool

| Judge 2 \ Judge 1 | 1 | 2 | 3 | 4 | 5 | 6 | 7 | 8 | 9 | 10 | 11 | NA | Total | % |
|---|---|---|---|---|---|---|---|---|---|---|---|---|---|---|
| 1 | 4 | | | | | | | | | | | | 6 | 67 |
| 2 | 1 | 5 | | | | | | | | | | | 7 | 86 |
| 3 | | | 3 | | | | | | | | | | 5 | 60 |
| 4 | | | 1 | 4 | | | | | | | | | 5 | 100 |
| 5 | | | | | 5 | | | | | | | | 5 | 100 |
| 6 | | | | | | 3 | | 1 | | | | | 4 | 100 |
| 7 | | | | | | | 4 | | | | | | 4 | 100 |
| 8 | | | | | | | | 3 | | | | | 4 | 75 |
| 9 | | | | | | | | | 3 | | | | 4 | 75 |
| 10 | | | | | | | | | | 3 | | | 5 | 60 |
| 11 | | | | | | | | | | | 3 | | 5 | 60 |
| NA | | | | | | | | | | | 3 | | | |

Total items placement: 54        Number of agreement: 43        Agreement ratio: 80 %

1. IT infrastructure flexibility
2. IT expertise
3. IOS use for integration
4. IOS use for communication
5. IOS use for intelligence
6. Collectivism
7. Long term orientation
8. Power symmetry
9. Uncertainty avoidance
10. Credibility
11. Benevolence

**Table 6.3** Items placement ratios: the first sorting round, the first pool

| Theoretical categories | Actual category | | | | | | | | | | | NA | Total | % |
|---|---|---|---|---|---|---|---|---|---|---|---|---|---|---|
| | 1 | 2 | 3 | 4 | 5 | 6 | 7 | 8 | 9 | 10 | 11 | | | |
| 1 | 9 | 1 | 1 | | 1 | | | | | | | | 12 | 75 |
| 2 | 2 | 11 | 1 | | | | | | | | | | 14 | 79 |
| 3 | 1 | | 8 | 1 | | | | | | | | | 10 | 80 |
| 4 | | | 2 | 8 | | | | | | | | | 10 | 80 |
| 5 | | | | | 10 | | | | | | | | 10 | 100 |
| 6 | | | | | | 6 | | 1 | 1 | | | | 8 | 75 |
| 7 | | | | | | | 8 | | | | | | 8 | 100 |
| 8 | | | | | | 1 | | 7 | | | | | 8 | 88 |
| 9 | | | | | | | 1 | | 7 | | | | 8 | 88 |
| 10 | | | | | | | | | | 8 | 2 | | 10 | 80 |
| 11 | | | | | | | | 1 | | 1 | 8 | | 10 | 80 |

Total items placement: 108          Number of hits: 90          Overall hit ratio: 83 %

1. IT infrastructure flexibility
2. IT expertise
3. IOS Use for Integration
4. IOS use for communication
5. IOS use for intelligence
6. Collectivism
7. Long term orientation
8. Power symmetry
9. Uncertainty avoidance
10. Credibility
11. Benevolence

**Table 6.4** Inter-judge raw agreement scores: the first sorting round, the second pool

| | Judge 1 | | | | | | | | | | | | | | |
|---|---|---|---|---|---|---|---|---|---|---|---|---|---|---|---|
| Judge 2 | 1 | 2 | 3 | 4 | 5 | 6 | 7 | 8 | 9 | 10 | 11 | 12 | NA | Total | % |
| 1 | 4 | | | | | | | | | | | | | 6 | 67 |
| 2 | | 5 | | | | | | | | | | | | 6 | 83 |
| 3 | | | 3 | | | | 1 | | | | | | | 6 | 50 |
| 4 | | 1 | | 4 | | | | | | | | | | 6 | 67 |
| 5 | | | | | 6 | | | | | | | | | 6 | 100 |
| 6 | 1 | | | | | 5 | | | | | | | | 6 | 83 |
| 7 | | | | | | | 3 | | | | | | | 6 | 50 |
| 8 | | | | | | | | 3 | | | | | | 4 | 75 |
| 9 | | | | | | | | 1 | 3 | | | | | 4 | 75 |
| 10 | | | | | | | | | | 3 | | | | 4 | 75 |
| 11 | | | | | | | | | | | 4 | | | 5 | 80 |
| 12 | | | | | | | | | | | | 3 | | 4 | 75 |
| NA | | | | | | | | | | | | | 0 | | |

Total items placement: 63                    Number of agreement: 51                    Agreement ratio: 81 %

1. Information sharing
2. Goal congruence
3. Decision synchronization
4. Incentive alignment
5. Resource sharing
6. Collaborative communication
7. Joint knowledge creation
8. Process efficiency
9. Offering flexibility
10. Business synergy
11. Quality
12. Innovation

**Table 6.5** Items placement ratios: the first sorting round, the second pool

| Theoretical categories | Actual category | | | | | | | | | | | | NA | Total | % |
|---|---|---|---|---|---|---|---|---|---|---|---|---|---|---|---|
| | 1 | 2 | 3 | 4 | 5 | 6 | 7 | 8 | 9 | 10 | 11 | 12 | NA | Total | % |
| 1 | 9 | | | | 1 | 1 | 1 | | | | | | | 12 | 75 |
| 2 | | 10 | | 1 | | 1 | | | | | | | | 12 | 83 |
| 3 | | | 8 | 1 | 1 | | 2 | | | | | | | 12 | 67 |
| 4 | | 2 | 1 | 9 | | | | | | | | | | 12 | 75 |
| 5 | | | | | 12 | | | | | | | | | 12 | 100 |
| 6 | 2 | | | | | 10 | | | | | | | | 12 | 83 |
| 7 | 1 | | 2 | | | | 9 | | | | | | | 12 | 75 |
| 8 | | | | | | | | 7 | | | 1 | | | 8 | 88 |
| 9 | | | | | | | | 2 | 6 | | | | | 8 | 75 |
| 10 | | | | | | | | 1 | | 7 | | | | 8 | 88 |
| 11 | | | | | | | | | | | 10 | | | 10 | 100 |
| 12 | | | | | | | | 2 | | | | 6 | | 8 | 75 |

Total items placement: 126                     Number of hits: 103                     Overall hit ratio: 82 %

1. Information sharing
2. Goal congruence
3. Decision synchronization
4. Incentive alignment
5. Resource sharing
6. Collaborative communication
7. Joint knowledge creation
8. Process efficiency
9. Offering flexibility
10. Business synergy
11. Quality
12. Innovation

**Table 6.6** Inter-judge raw agreement scores: the second sorting round, the first pool

| Judge 2 | Judge 1 | | | | | | | | | | | | | |
|---|---|---|---|---|---|---|---|---|---|---|---|---|---|---|
| | 1 | 2 | 3 | 4 | 5 | 6 | 7 | 8 | 9 | 10 | 11 | NA | Total | % |
| 1 | 5 | | | | | | | | | | | | 5 | 100 |
| 2 | | 4 | | | | | | | | | | | 6 | 67 |
| 3 | | | 4 | | | | | | | | | | 5 | 80 |
| 4 | | | | 4 | | | | | | | | | 5 | 80 |
| 5 | | | | | 5 | | | | | | | | 5 | 100 |
| 6 | | | | | | 3 | | | | | | | 4 | 75 |
| 7 | | | | | | | 4 | | | | | | 4 | 100 |
| 8 | | | | | | | | 3 | | | | | 4 | 75 |
| 9 | | | | | | | | | 3 | | | | 4 | 75 |
| 10 | | | | | | | | | | 4 | | | 4 | 75 |
| 11 | | | | | | | | | | | 4 | | 5 | 80 |
| NA | | | | | | | | | | | | | 5 | 80 |

Total items placement: 52          Number of agreement: 43          Agreement ratio: 83 %

1. IT infrastructure flexibility
2. IT expertise
3. IOS use for integration
4. IOS use for communication
5. IOS Use for intelligence
6. Collectivism
7. Long term orientation
8. Power symmetry
9. Uncertainty avoidance
10. Credibility
11. Benevolence

**Table 6.7** Items placement ratios: the second sorting round, the first pool

| Theoretical categories | Actual category | | | | | | | | | | | NA | Total | % |
|---|---|---|---|---|---|---|---|---|---|---|---|---|---|---|
| | 1 | 2 | 3 | 4 | 5 | 6 | 7 | 8 | 9 | 10 | 11 | | | |
| 1 | 10 | | | | | | | | | | | | 10 | 100 |
| 2 | 2 | 10 | | | | | | | | | | | 12 | 83 |
| 3 | | | 9 | | | | | | | | | | 10 | 90 |
| 4 | | | 1 | 9 | | | | | | | | | 10 | 90 |
| 5 | | | | | 10 | | | | | | | | 10 | 100 |
| 6 | | | | | | 7 | 1 | | | | | | 8 | 88 |
| 7 | | | | | | | 8 | | | | | | 8 | 100 |
| 8 | | | | | | 1 | | 7 | | | | | 8 | 88 |
| 9 | | | | | | 1 | | | 7 | | | | 8 | 88 |
| 10 | | | | | | | | | | 9 | 1 | | 10 | 90 |
| 11 | | | | | | | | | | 1 | 9 | | 10 | 90 |

Total items placement: 104          Number of hits: 95          Overall hit ratio: 91 %

1. IT infrastructure flexibility
2. IT expertise
3. IOS use for integration
4. IOS use for communication
5. IOS use for intelligence
6. Collectivism
7. Long term orientation
8. Power symmetry
9. Uncertainty avoidance
10. Credibility
11. Benevolence

**Table 6.8** Inter-judge raw agreement scores: the second sorting round, the second pool

| Judge 2 | Judge 1 | | | | | | | | | | | | NA | Total | % |
|---|---|---|---|---|---|---|---|---|---|---|---|---|---|---|---|
| | 1 | 2 | 3 | 4 | 5 | 6 | 7 | 8 | 9 | 10 | 11 | 12 | | | |
| 1 | 4 | | | | | | | | | | | | | 5 | 80 |
| 2 | | 4 | | | | | | | | | | | | 5 | 80 |
| 3 | | | 4 | | | | | | | | | | | 5 | 80 |
| 4 | | | | 4 | | | | | | | | | | 5 | 80 |
| 5 | | | | | 4 | | | | | | | | | 5 | 80 |
| 6 | | | | | | 5 | | | | | | | | 5 | 100 |
| 7 | | | | | | | 4 | | | | | | | 5 | 80 |
| 8 | | | | | | | | 3 | | | | | | 4 | 75 |
| 9 | | | | | | | | | 4 | | | | | 4 | 100 |
| 10 | | | | | | | | | | 3 | | | | 4 | 75 |
| 11 | | | | | | | | | | | 4 | | | 4 | 100 |
| 12 | | | | | | | | | | | | 4 | | 4 | 100 |
| NA | | | | | | | | | | | | | | | |

Total items placement: 55        Number of agreement: 47        Agreement ratio: 85 %

1. Information sharing
2. Goal congruence
3. Decision synchronization
4. Incentive alignment
5. Resource sharing
6. Collaborative communication
7. Joint knowledge creation
8. Process efficiency
9. Offering flexibility
10. Business synergy
11. Quality
12. Innovation

**Table 6.9** Items placement ratios: the second sorting round, the second pool

| Theoretical categories | Actual category | | | | | | | | | | | | | | |
|---|---|---|---|---|---|---|---|---|---|---|---|---|---|---|---|
| | 1 | 2 | 3 | 4 | 5 | 6 | 7 | 8 | 9 | 10 | 11 | 12 | NA | Total | % |
| 1 | 9 | | | | | 1 | | | | | | | | 10 | 90 |
| 2 | | 9 | | 1 | | | | | | | | | | 10 | 90 |
| 3 | | | 8 | | 1 | | | | | | | | | 10 | 80 |
| 4 | | | 1 | 9 | | | | | | | | | | 10 | 90 |
| 5 | | | 1 | | 9 | | | | | | | | | 10 | 90 |
| 6 | | | | | | 10 | | | | | | | | 10 | 100 |
| 7 | | | 1 | | | | 9 | | | | | | | 10 | 90 |
| 8 | | | | | | | | 7 | | | 1 | | | 8 | 88 |
| 9 | | | | | | | | | 8 | | | | | 8 | 100 |
| 10 | | | | | | | | 1 | | 7 | | | | 8 | 88 |
| 11 | | | | | | | | | | | 8 | | | 8 | 100 |
| 12 | | | | | | | | 2 | | | | 6 | | 8 | 75 |

Total items placement: 110                Number of hits: 99                Overall hit ratio: 90 %

1. Information sharing
2. Goal congruence
3. Decision synchronization
4. Incentive alignment
5. Resource sharing
6. Collaborative communication
7. Joint knowledge creation
8. Process efficiency
9. Offering flexibility
10. Business synergy
11. Quality
12. Innovation

**Table 6.10** Inter-judge agreements

| Agreement measure | Round 1 | Round 2 |
|---|---|---|
| First pool | | |
| Raw agreement | 80 % | 83 % |
| Cohen's Kappa | 0.78 | 0.82 |
| Placement ratio summary | | |
| IT infrastructure flexibility | 75 % | 100 % |
| IT expertise | 79 % | 83 % |
| IOS use for integration | 80 % | 90 % |
| IOS use for communication | 80 % | 90 % |
| IOS use for intelligence | 100 % | 100 % |
| Collectivism | 75 % | 88 % |
| Long term orientation | 100 % | 100 % |
| Power symmetry | 88 % | 88 % |
| Uncertainty avoidance | 88 % | 88 % |
| Credibility | 80 % | 90 % |
| Benevolence | 80 % | 90 % |
| Average | 83 % | 91 % |
| Second pool | | |
| Raw agreement | 81 % | 85 % |
| Cohen's Kappa | 0.80 | 0.84 |
| Placement ratio summary | | |
| Information sharing | 75 % | 90 % |
| Goal congruence | 83 % | 90 % |
| Decision synchronization | 67 % | 80 % |
| Incentive alignment | 75 % | 90 % |
| Resource sharing | 100 % | 90 % |
| Collaborative communication | 83 % | 100 % |
| Joint knowledge creation | 75 % | 90 % |
| Process efficiency | 88 % | 88 % |
| Offering flexibility | 75 % | 100 % |
| Business synergy | 88 % | 88 % |
| Quality | 100 % | 100 % |
| Innovation | 75 % | 75 % |
| Average | 82 % | 90 % |

seven items deleted and ten reworded. Deleted and reworded items are noted in Appendix A and D respectively.

After deleting and rewording items from the first round, a second sorting round was conducted with another two judges. The results are shown in Tables 6.6–6.9. For the first pool, the inter-judge raw agreement scores average 83 % (Table 6.6), the overall placement ratio of items is 91 % (Table 6.7), and Kappa scores average 0.82 (Table 6.10). For the second pool, the inter-judge raw scores average 85 % (Table 6.8), the overall placement ratio of items is 90 % (Table 6.9), and Kappa scores average 0.84 (Table 6.10). Based on the guidelines of Landis and Koch (1977), the Kappa scores of 0.82 and 0.84 in the two pools respectively indicate an excellent level of agreement.

**Table 6.11** Constructs, sub-constructs, and number of items

| Constructs | Sub-constructs | # of items |
|---|---|---|
| IT resources | IT infrastructure flexibility | 5 |
| | IT expertise | 6 |
| IOS appropriation | IOS use for integration | 5 |
| | IOS use for communication | 5 |
| | IOS use for intelligence | 5 |
| Collaborative culture | Collectivism | 4 |
| | Long term orientation | 4 |
| | Power symmetry | 4 |
| | Uncertainty avoidance | 4 |
| Trust | Credibility | 5 |
| | Benevolence | 5 |
| Supply chain collaboration | Information sharing | 5 |
| | Goal congruence | 5 |
| | Decision synchronization | 5 |
| | Resource sharing | 5 |
| | Incentive Alignment | 5 |
| | Collaborative communication | 5 |
| | Joint knowledge creation | 5 |
| Collaborative advantage | Process efficiency | 4 |
| | Offering flexibility | 4 |
| | Business synergy | 4 |
| | Quality | 4 |
| | Innovation | 4 |
| Firm performance | | 7 |
| Total | | 114 |

After two rounds of Q-Sort, 107 items were distributed to six academicians who reviewed each item and indicated to keep, drop, modify, or add items to the constructs. The purpose was to further refine the items and assess whether the items measured the proposed sub-constructs that they were supposed to measure based on the definitions provided, or whether any additional items were needed to cover the domain. Based on the feedback from the reviewers, items were further modified. Overall, 114 (107 + 7) questionnaire items, including seven items adapted from Li (2002) for the construct of firm performance, were sent out for a large-scale survey (Table 6.11). The large-scale survey questionnaire items are provided in Appendices D and E.

# References

Angeles, R., & Nath, R. (2001). Partner congruence in electronic data interchange (EDI) enabled relationships. *Journal of Business Logistics, 22*(2), 109–127.

Angeles, R., & Nath, R. (2003). Electronic supply chain partnerships: Reconsidering relationship attributes in customer-supplier dyads. *Information Resources Management Journal, 16*(3), 59–84.

Armstrong, C. P., & Sambamurthy, V. (1999). Information technology assimilation in firms: The influence of senior leadership and IT infrastructures. *Information Systems Research, 10*(4), 304–327.

Ba, S., & Pavlou, P. A. (2002). Evidence of the effect of trust building technology in electronic markets: Price premiums and buyer behavior. *MIS Quarterly, 26*(3), 243–268.

Bafoutsou, G., & Mentzas, G. (2002). Review and functional classification of collaborative systems. *International Journal of Information Management, 22*(4), 281–306.

Bagchi, P. K., & Skjoett-Larsen, T. (2005). Supply chain integration: A survey. *The International Journal of Logistics Management, 16*(2), 275–294.

Barua, A., Konana, P., Whinston, A. B., & Yin, F. (2004). An empirical investigation of net-enabled business value. *MIS Quarterly, 28*(4), 585–620.

Bates, K. A., Amundson, S. D., Schroeder, R. G., & Morris, W. T. (1995). The crucial interrelationship between manufacturing strategy and organizational culture. *Management Science, 41*(10), 1565–1580.

Bensaou, M., & Venkatraman, N. (1995). Configurations of interorganizational relationships: A comparison between U.S. and Japanese automakers. *Management Science, 41*(9), 1471–1492.

Bharadwaj, A. S. (2000). A resource based perspective on information technology capability and firm performance: An empirical investigation. *MIS Quarterly, 24*(1), 169–196.

Bhatt, G. D., & Grover, V. (2005). Types of information technology capabilities and their role in competitive advantage: An empirical study. *Journal of Management Information Systems, 22*(2), 253–277.

Bhattacharya, R., Devinney, T. M., & Pillutla, M. M. (1998). A formal model of trust based on outcomes. *Academy of Management Review, 23*(3), 459–472.

Boddy, D., Macbeth, D., & Wagner, B. (2000). Implementing collaboration between organizations: An empirical study of supply chain partnering. *Journal of Management Studies, 37*(7), 1003–1019.

Bowersox, D. J., Closs, D. J., & Stank, T. P. (2003). How to master cross-enterprise collaboration. *Supply Chain Management Review, 7*(4), 18–27.

Burnes, B., & New, S. (1996). Understanding supply chain improvement. *European Journal of Purchasing and Supply Management, 2*(1), 21–30.

Byrd, T. A., & Turner, D. E. (2000). Measuring the flexibility of information technology infrastructure: Exploratory analysis of a construct. *Journal of Management Information Systems, 17*(1), 167–208.

Chi, L., & Holsapple, C. W. (2005). Understanding computer-mediated interorganizational collaboration: A model and framework. *Journal of Knowledge Management, 9*(1), 53–75.

Christiaanse, E., & Venkatraman, N. (2002). Beyond SABRE: An empirical test of expertise exploitation in electronic channels. *MIS Quarterly, 26*(1), 15–38.

Churchill, G. A. (1979). A paradigm for developing better measures of marketing constructs. *Journal of Marketing Research, 16*, 64–73.

Clark, K. B., & Fujimoto, T. (1991). *Product development performance: Strategy, organization, and management in the world auto industry.* Boston, MA: Harvard Business School Press.

Cooper, C., Lambert, C., & Pagh, D. (1997a). Supply chain management: More than a new name for logistics. *The International Journal of Logistics Management, 8*(1), 1–14.

Cooper, M., Ellram, L. M., Gardner, J. T., & Hanks, A. M. (1997b). Meshing multiple alliances. *Journal of Business Logistics, 18*(1), 67–89.

Das, T. K., & Teng, B. S. (1998). Between trust and control: developing confidence in partner cooperation in alliances. *Academy of Management Review, 23*(3), 491–512.

De Wever, S., Martens, R., & Vandenbempt, K. (2005). The impact of trust on strategic resource acquisition through interorganizational networks: Towards a conceptual model. *Human Relations, 58*(12), 1523–1543.

Dehning, B., & Richardson V. J. (2002). Returns on investments in information technology: A research synthesis. *Journal of Information Systems, 16*(1), 7–30.

Dyer, J. H. (1996). Does governance matter? Keiretsu alliances and asset specificity as sources of Japanese competitive advantage. *Organization Science, 7*(6), 649–666.

Dyer, J. H., & Singh, H. (1998). The relational view: Cooperative strategy and sources of interorganizational competitive advantage. *Academy of Management Review, 23*(4), 660–679.

Ellram, L. M. (1995). A managerial guideline for the development and implementation of purchasing partnerships. *International Journal of Purchasing and Materials Management, 31*(2), 9–16.

Ellram, L. M., & Edis, O. R. V. (1996). A case study of successful partnering implementation. *International Journal of Purchasing and Materials Management, 32*(2), 20–28.

Ferratt, T. W., Lederer, A. L., Hall, J. M., & Krella, J. M. (1996). Swords and plowshares: Information technology for collaborative advantage. *Information and Management, 30*(3), 131–142.

Fisher, M. L. (1997). What is the right supply chain for your product? *Harvard Business Review, 75*(2), 105–116.

Golicic, S. L., Foggin, J. H., & Mentzer, J. T. (2003). Relationship magnitude and its role in interorganizational relationship structure. *Journal of Business Logistics, 24*(1), 57–75.

Gosain, S., Malhotra, A., & El Sawy, O. A. (2004). Coordinating for flexibility in e-business supply chains. *Journal of Management Information Systems, 21*(3), 7–45.

Grieger, N. (2003). Electronic marketplaces: A literature review and a call for supply chain management research. *European Journal of Operational Research, 144*(2), 280–294.

Grover, V., Teng, J., & Fiedler, K. (2002). Investigating the role of information technology in building buyer-supplier relationships. *Journal of Association for Information Systems, 3*, 217–245.

Handfield, R. B., & Pannesi, R. T. (1995). Antecedents of leadtime competitiveness in make-to-order manufacturing firms. *International Journal of Production Research, 33*(2), 511–537.

Hardy, C., Phillips, N., & Lawrence, T. B. (2003). Resources, knowledge and influence: The organizational effects of interorganizational collaboration. *Journal of Management Studies, 40*(2), 321–347.

Harland, C. M., Zheng, J., Johnsen, T. E., & Lamming, R. C. (2004). A conceptual model for researching the creation and operation of supply networks. *British Journal of Management, 15*(1), 1–21.

Hofstede, G. (1980). *Culture's consequences: International differences in work-related values.* London: Sage Publications.

Hofstede, G. (1991). *Cultures and organizations: Software of the mind.* London, UK: McGraw-Hill.

Hofstede, G. (2001). *Culture's consequences* (2nd ed.). Thousand Oaks, CA: Sage Publications.

Jap, S. D. (2001). Perspectives on joint competitive advantages in buyer-supplier relationships. *International Journal of Research in Marketing, 18*(1/2), 19–35.

Johnson, J. J., & Sohi, R. S. (2003). The development of interfirm partnering competence: Platforms for learning, learning activities and consequences of learning. *Journal of Business Research, 56*(9), 757–766.

Johnson, M. E., & Whang, S. (2002). E-business and supply chain management: An overview and framework. *Production and Operations management, 11*(4), 413–422.

Johnson, D. A., McCutcheon, D. M., Stuart, F. I., & Kerwood, H. (2004). Effect of supplier trust on performance of cooperative supplier relationships. *Journal of Operations Management, 22*(1), 23–38.

Kanji, G. K., & Wong, A. (1999). Business excellence model for supply chain management. *Total Quality Management, 10*(8), 1147–1168.

Kanter, R. M. (1994). Collaborative advantage: The art of alliances. *Harvard Business Review*, 72, 96–108.

Kaufman, A., Wood, C. H., & Theyel, G. (2000). Collaboration and technology linkages: A strategic supplier typology. *Strategic Management Journal*, 21(6), 649–663.

Kock, N., & Nosek, J. (2005). Expanding the boundaries of e-collaboration. *IEEE Transactions on Professional Communication*, 48(1), 1–9.

Kumar, K., Van Dissel, H. G., & Bielli, P. (1998). The merchant of Prato revisited: Toward a third rationality of information systems. *MIS Quarterly*, 22(2), 199–226.

Lambert, D. M., Emmelhainz, M. A., & Gardner, J. T. (1996). Developing and implementing supply chain partnerships. *The International Journal of Logistics Management*, 7(2), 1–17.

Lambert, D. M., Emmelhainz, M. A., & Gardner, J. T. (1999). Building successful logistics partnerships. *Journal of Business Logistics*, 20(1), 165–181.

Landis, J. R., & Koch, G. G. (1977). The measurement of observer agreement for categorical data. *Biometrics*, 33, 159–174.

Lee, H. L., Padmanabhan, V., & Whang, S. (1997). The bullwhip effect in supply chain. *Sloan Management Review*, 38(3), 93–102.

Lejeune, N., & Yakova, N. (2005). On Characterizing the 4 C's in supply china management. *Journal of Operations Management*, 23(1), 81–100.

Li, S. (2002). *An integrated model for supply chain management practice, performance, and competitive advantage*. Unpublished dissertation, University of Toledo, Toledo.

Luo, X., Slotegraaf, R. J., & Pan, X. (2006). Cross-functional coopetition: The simultaneous role of cooperation and competition within firms. *Journal of Marketing*, 70(2), 67–80.

Macbeth, D. (1998). Partnering—why not? In *The Second Worldwide Research Symposium on Purchasing and Supply Chain Management*. London, England, 351–362.

Malhotra, A., Majchrzak, A., Carman, R., & Lott, V. (2001). Radical innovation without collocation: A case study and Boeing-Rocketdyne. *MIS Quarterly*, 25(2), 229–249.

Malhotra, A., Gasain, S., & El Sawy, O. A. (2005). Absorptive capacity configurations in supply chains: Gearing for partner-enabled market knowledge creation. *MIS Quarterly*, 29(1), 145–187.

Manthou, V., Vlachopoulou, M., & Folinas, D. (2004). Virtual e-Chain (VeC) model for supply chain collaboration. *International Journal of Production Economics*, 87(3), 241–250.

Marquez, A. C., Bianchi, C., & Gupta, J. N. D. (2004). Operational and financial effectiveness of e-collaboration tools in supply chain integration. *European Journal of Operational Research*, 159(2), 348–363.

McDonnell, M. (2001). E-collaboration: Transforming your supply chain into a dynamic trading community. *Supply Chain Practice*, 3(2), 80–89.

McKnight, D. H., & Chervany, N. L. (2002). What trust means in e-commerce customer relationships: An interdisciplinary conceptual typology. *International Journal of Electronic Commerce*, 6(2), 35–53.

Mehra, A., & Nissen, M. (1998). Case study: Intelligent software supply chain agents using ADE. In J. Baxter & B. Logan (Eds.), *Proceedings of the Workshop on Software Tools for Developing Agents* (pp. 53–62). Madison, WI: American Association for Artificial Intelligence.

Melville, N., Kraemer, K., & Gurbaxani, V. (2004). Information technology and organizational performance: An integrative model of it business value. *MIS Quarterly*, 28(2), 283–322.

Mentzer, J. T., DeWitt, W., Keebler, J. S., Min, S., Nix, N. W., Smith, C. D., et al. (2001). Defining supply chain management. *Journal of Business Logistics*, 22(2), 1–25.

Milton, N., Shadbolt, N., Cottam, H., & Hammersley, M. (1999). Towards a knowledge technology for knowledge management. *International Journal of Human-Computer Studies*, 51(3), 615–641.

Min, S., Roath, A., Daugherty, P. J., Genchev, S. E., Chen, H., & Arndt, A. D. (2005). Supply chain collaboration: What's happening? *International Journal of Logistics Management*, 16(2), 237–256.

Mohr, J., & Nevin, J. R. (1990). Communication strategies in marketing channels: A theoretical perspective. *Journal of Marketing*, 54(4), 36–51.

Moore, G. C., & Benbasat, I. (1991). Development of an instrument to measure the perceptions of adopting an information technology innovation. *Information Systems Research, 2*(3), 192–222.

Mukhopadhyay, T., & Kekre, S. (2002). Strategic and operational benefits of electronic integration in B2B procurement process. *Management Science, 48*(10), 1301–1313.

Narayandas, D., & Rangan, K. V. (2004). Building and sustaining buyer-seller relationships in mature industrial market. *Journal of Marketing, 68*(3), 63–77.

Nesheim, T. (2001). Externalization of the core: Antecedents of collaborative relationships with suppliers. *European Journal of Purchasing and Supply Management, 7*(4), 217–225.

Nissen, M. E., & Sengupta, K. (2006). Incorporating software agents into supply chains: Experimental investigation with a procurement task. *MIS Quarterly, 30*(1), 145–166.

Nooteboom, B., Berger, H., & Noorderhaven, N. G. (1997). Effects of trust and governance on relational risk. *Academy of Management Journal, 40*(2), 308–338.

Pavlou, P. A. (2002a). Institution-based trust in interorganizational exchange relationships: The role of online B2B marketplaces on trust formation. *Journal of Strategic Information Systems, 11*(3/4), 215–243.

Pavlou, P. A. (2002b). Trustworthiness as a source of competitive advantage in online auction markets. In D. H. Nagao (Ed.), *Best Paper Proceedings of the Academy of Management Conference.* Denver, CO, A1–A6.

Pavlou, P. A., & Gefen, D. (2004). Building effective online marketplaces with institution-based trust. *Information Systems Research, 15*(1), 35–62.

Peppard, J., & Ward, J. (2004). Beyond strategic information systems: Towards an IS capability. *Journal of Strategic Information Systems, 13*(2), 167–194.

Piccoli, G., & Ives, B. (2005). IT-dependent strategic initiatives and sustained competitive advantage: A review and synthesis of the literature. *MIS Quarterly, 29*(4), 747–776.

Poirier, C. C., & Houser, W. F. (1993). *Business partnering for continuous improvement.* San Francisco, CA: Berrett-Koehler, 56, 201.

Prahinski, C., & Benton, W. C. (2004). Supplier evaluations: Communication strategies to improve supplier performance. *Journal of Operations Management, 22*(1), 39–62.

Ranganathan, C., Dhaliwal, J. S., & Teo, T. S. (2004). Assimilation and diffusion of web technologies in supply-chain management: An examination of key drivers and performance impacts. *International Journal of Electronic Commerce, 9*(1), 127–161.

Ravichandran, T., & Lertwongsatien, C. (2005). Effect of information systems resources and capabilities on firm performance: A resource-based perspective. *Journal of Management Information Systems, 21*(4), 237–276.

Ray, G., Muhanna, W. A., & Barney, J. B. (2005). Information technology and the performance of the customer service process: A resource-based analysis. *MIS Quarterly, 29*(4), 625–652.

Ring, P. S., & Van de Ven, A. H. (1992). Structuring cooperative relationships between organizations. *Strategic Management Journal, 13*(7), 483–498.

Ross, J. W., Beath, C. M., & Goodhue, D. L. (1996). Develop long-term competitiveness through IT assets. *Sloan Management Review, 38*(1), 31–42.

Saeed, K. A., Malhotra, M. K., & Grover, V. (2005). Examining the impact of interorganizational systems on process efficiency and sourcing leverage in buyer–supplier dyads. *Decision Sciences, 36*(3), 365–396.

Salisbury, W. D., Chin, W. W., Gopal, A., & Newsted, P. R. (2002). Research report: Better theory through measurement—developing a scale to capture consensus on appropriation. *Information Systems Research, 13*(1), 91–103.

Scheer, L., Kumar, N., & Steenkamp, J. (2003). Reactions to perceived inequality in U.S. and Dutch interorganizational relationships. *Academy of Management Journal, 46*(3), 303–316.

Segars, A. H., & Grover, V. (1998). Strategic information systems planning success: An investigation of the construct and its measurement. *MIS Quarterly, 22*, 139–163.

Sheu, C., Yen, H. R., & Chae, D. (2006). Determinants of supplier-retailer collaboration: Evidence from an international study. *International Journal of Operations and Production Management, 26*(1), 24–49.

Simatupang, T. M., & Sridharan, R. (2005a). An integrative framework for supply chain collaboration. *International Journal of Logistics Management, 16*(2), 257–274.

Simatupang, T. M., & Sridharan, R. (2005b). Supply chain discontent. *Business Process Management Journal, 11*(4), 349–369.

Simatupang, T. M., & Sridharan, R. (2005c). The collaboration index: A measure for supply chain collaboration. *International Journal of Physical Distribution and Logistics Management, 35*(1), 44–62.

Son, J., Narasimhan, S., & Riggins, F. J. (2005). Effects of relational factors and channel climate on EDI usage in the customer-supplier relationship. *Journal of Management information Systems, 22*(1), 321–353.

Sriram, V., Krapfel, R., & Spekman, R. E. (1992). Antecedents to buyer-seller collaboration: An analysis from the buyer's perspective. *Journal of Business Research, 25*(4), 303–320.

Stank, T. P., Keller, S. B., & Daugherty, P. J. (2001). Supply chain collaboration and logistical service performance. *Journal of Business Logistics, 22*(1), 29–48.

Steensma, H. K., Marino, L., Weaver, K. M., & Dickson, P. H. (2000). The influence of national culture on the formation of technology alliances by entrepreneurial firms. *Academy of Management Journal, 43*(5), 951–973.

Subramani, M. (2004). How do suppliers benefit from information technology use in supply chain relationships? *MIS Quarterly, 28*(1), 45–73.

Tanriverdi, H. (2006). Performance effects of information technology synergies in multibusiness firms. *MIS Quarterly, 30*(1), 57–77.

Tuten, T. L., & Urban, D. J. (2001). An expanded model of business-to-business partnership foundation and success. *Industrial Marketing Management, 30*(2), 149–164.

Vangen, S., & Huxham, C. (2003). Enacting leadership for collaborative advantage: Dilemmas of ideology and pragmatism in the activities of partnership managers. *British Journal of Management, 14*(Supplement 1), S61–S76.

Verwaal, E., & Hesselmans, M. (2004). Drivers of supply network governance: An explorative study of the Dutch chemical industry. *European Management Journal, 22*(4), 442–451.

Weill, P., Broadbent, M., & St. Clair, D. (1996). IT value and the role of IT infrastructure investments. In J. N. Luftman (Ed.), *Competing in the information age: Strategic alignment in practice* (pp. 361–384). New York: Oxford University Press.

Wuyts, S., & Geyskens, I. (2005). The formation of buyer–supplier relationships: Detailed contract drafting and close partner selection. *Journal of Marketing, 69*(4), 103–117.

Zaheer, A., McEvily, B., & Perrone, V. (1998). Does trust matter? Exploring the effects of interorganizational and interpersonal trust on performance. *Organization Science, 9*(2), 141–159.

Zhu, K. (2004). The complementarity of information technology infrastructure and e-commerce capability: A resource-based assessment of their business value. *Journal of Management Information Systems, 21*(1), 167–202.

# Chapter 7
# Large-Scale Analysis and Testing

**Abstract** Collecting large-scale data and testing the hypotheses in the framework are challenging. In this chapter, we describe the large-scale web survey method and present the analysis procedure and results. Data collection process includes sampling design, purchasing executive email list, web survey design, and managing waves of survey. Confirmatory factor analysis is used to assess unidimensionality, reliability, convergent validity, discriminant validity, and second-order construct validity. A LISREL structural model is used to test the hypotheses. The results are reported and discussed. To further purify the items and assess unidimensionality, reliability, convergent, and discriminant validity, a large-scale Web survey was conducted. The main analysis tool used is the confirmatory factor analysis with structural equation modeling (SEM).

## 7.1 Sampling Design

The sample respondents were expected to have knowledge or experience in supply chain management and information systems use, as well as general knowledge about the supply chain performance and firm's performance indicators. The target respondents were CEOs, presidents, vice presidents, directors, or managers in the manufacturing firms across the U.S., whose job responsibilities were in the areas of purchasing/procurement, manufacturing/operations, distribution/warehouse, transportation/logistics, supply chain management, and/or information technology. The respondents were expected to be the representatives of different supply chain tiers (e.g., raw material suppliers, component suppliers, assemblers, manufacturers, wholesalers, distributors, and retailers) and different firm sizes to achieve greater generalizability. The sample respondents were expected to cover the following seven Standard Industrial Classification (SIC) codes (Table 7.1).

An email list of 5,000 target respondents were purchased from Council of Supply Chain Management Professionals (CSCMP), a prestigious association of professionals in the area of supply chain management from diverse industries

M. Cao and Q. Zhang, *Supply Chain Collaboration*,
DOI: 10.1007/978-1-4471-4591-2_7, © Springer-Verlag London 2013

**Table 7.1** Standard industrial classification codes

| | |
|---|---|
| Furniture and fixtures | SIC 25 |
| Rubber and plastic products | SIC 30 |
| Fabricated metal products | SIC 34 |
| Industrial machinery and equipment | SIC 35 |
| Electric and electronic equipment | SIC 36 |
| Transportation equipment | SIC 37 |
| Instruments and related products | SIC 38 |

across the U.S., and lead411.com, a professional list company which specializes in providing executive level email lists. The survey was administered online because the Internet not only increases the richness of information but also increases the reach of information (Laudon and Laudon 2004). The purpose of using Web survey is to reach as many respondents as possible and retrieve as much information as possible in short time (Crawford et al. 2002).

The email list was refined to eliminate multiple names from the same organization. The person with the most relevant job title was picked and the others were removed. In this process, 249 names were removed from the email list. An invitation to participate in the survey, which explained the purpose of study, the instructions for completing the questionnaire, and measures to securely handle the data collected, were sent via email to 4,751 potential respondents. For the convenience of the respondents, three options were provided to complete and submit the questionnaire: (1) On-line: Click on the link that would take the respondents to the survey website to complete the survey; (2) Fax: Click on the link that would allow the respondents to download a copy of the questionnaire and send it by fax; (3) Regular mail: Email back to request a hard copy to be sent through regular mail and return it through either fax or regular mail.

After the first wave of emails was sent, the researcher did the second refinement of the email list by removing the names from the following emails: (1) emails that were undeliverable, (2) returned emails saying that target respondents were no longer with the company, (3) returned emails saying that target respondents did not work in the supply chain area, (4) returned emails saying that target respondents refused to participate because of time pressure or organization policy, or they felt they were not qualified to provide the answers. The refinement resulted in the removal of another 1,213 names. Therefore, the actual mailing list contained 3,538 names.

To improve the response rate, three waves of emails were sent once a week. A total of 152 responses were obtained on-line after the first wave of emails. The second and third wave generated 71 (69 on-line, 2 via mail) and 4 (2 on-line, 2 via mail) responses respectively. Out of the 227 responses received (16 incomplete), 211 are usable resulting in a response rate of 6.0 % (211/3538). Based on the information collected on the website, the number of unique clicks (one click per IP address is counted) is 702 resulting in a click through rate of 19.8 % (702/3538). The response rate out of the unique clicks is 30.1 % (211/702).

**Table 7.2** Comparison of first-wave and second/third-wave respondents

| Variables | First-wave frequency | Second/third wave (expected frequency $f_e$) | Second/third wave (observed frequency $f_o$) | Chi-square test |
|---|---|---|---|---|
| **SIC** | | | | |
| 25 | 4 | 2 | 4 | $\chi^2=10.00$ |
| 30 | 4 | 2 | 5 | df = 7 |
| 34 | 25 | 13 | 15 | p = 0.17 |
| 35 | 23 | 12 | 7 | |
| 36 | 43 | 22 | 24 | |
| 37 | 21 | 11 | 8 | |
| 38 | 15 | 8 | 8 | |
| Others | 4 | 2 | 1 | |
| **Firm size** | | | | |
| 1–50 | 7 | 4 | 3 | $\chi^2=4.71$ |
| 51–100 | 12 | 6 | 4 | df = 5 |
| 101–250 | 27 | 14 | 11 | p = 0.45 |
| 251–500 | 34 | 18 | 24 | |
| 501–1000 | 8 | 4 | 6 | |
| 1001+ | 51 | 26 | 24 | |
| **Job title** | | | | |
| CEO/President | 36 | 19 | 18 | $\chi^2=5.48$ |
| Vice president | 62 | 32 | 39 | df = 4 |
| Manager | 20 | 10 | 7 | p = 0.24 |
| Director | 17 | 9 | 6 | |
| Others | 4 | 2 | 4 | |

Sample characteristics appear on Table 7.2 based on SIC code, firm size, and respondents' job titles. The respondents come from manufacturing industries, namely, SIC 25, 30, 34, 35, 36, 37 and 38. The highest four respondent categories by SIC code are 34, 35, 36, and 37 (i.e., 79 % of respondents). About 80 % of firms have 100–500 or 1001 and more employees. 73 % of the respondents are presidents/CEO & vice presidents; 24 % are managers and directors.

A Chi-square test is conducted to check non-response bias. The results (see Table 7.2) show that there is no significant difference between the first-wave and second/third-wave respondents by all three categories (i.e., SIC code, firm size, and job title) at the level of 0.1. It exhibits that received questionnaires from respondents represent an unbiased sample.

## 7.2 Large-Scale Measurement Analysis

### 7.2.1 Large-Scale Measurement Methods

Using confirmatory factor analysis with LISREL, steps were undertaken tocheck (1) unidimensionality and convergent validity, (2) reliability, (3) discriminant validity, and (4) second-order construct validity of the measurement. Unidimensionality is defined as the existence of a latent construct underlying a set of measures. Convergent validity is an assessment of the consistency in measurements across multiple operationalizations. Unidimensionality is assessed by the fit indices of one-dimensional model for each construct and convergent validity is assessed by the significance of t-values of each measurement indicator.

Based on an evaluation of the fit of a one-dimensional model for each construct, iterative modification were undertaken in the spirit of a specification search, i.e., modifications were made to drop items with loadings less than 0.7 or items with high correlated errors to improve model fit (Hair et al. 1995). In all cases where refinement was indicated, items were deleted if such action was theoretically sound (Anderson 1987), and the deletions were done one at each step (Segars and Grover 1993; Hair et al. 1995). Model modifications were continued until all parameter estimates and model fits were judged to be satisfactory.

The overall model fit can be tested using the comparative fit index (CFI), non-normed fit index (NNFI), root mean square error of approximation (RMSEA), and normed Chi-square (i.e., $\chi^2$ per degree of freedom) (Bentler and Bonnet 1980; Byrne 1989; Bentler 1990; Hair et al. 1995; Chau 1997; Heck 1998). Values of CFI and NNFI between 0.80 and 0.89 represent a reasonable fit (Segars and Grover 1993) and scores of 0.90 or higher are evidence of good fit (Byrne 1989; Joreskog and Sorbom 1986; Papke-Shields et al. 2002). Values of RMSEA less than 0.08 are acceptable (Hair et al. 1995; Joreskog and Sorbom 1986). The normed Chi-square ($\chi^2$ divided by degrees of freedom (d.f.)) estimates the relative efficiency of competing models. For this statistic, a value less than 3.0 indicates a good fit (Segars and Grover 1993; Papke-Shields et al. 2002).

The typical approach to reliability assessment is the Cronbach's $\alpha$ coefficient. However, Cronbach's $\alpha$ is based on the restricted assumption of equal importance of all indicators. Following Hair et al. (1995), the composite reliability ($\rho_c$) and the average variance extracted (AVE) of multiple indicators of a construct can be used to assess reliability of a construct. The formulas for calculating them are shown below. When AVE is greater than 0.50 and $\rho_c$ is greater than 0.70, it implies that the variance by the trait is more than that by error components (Hair et al. 1995).

$$\rho_c = \frac{\left(\sum \lambda_i\right)^2}{\left(\sum \lambda_i\right)^2 + \sum \epsilon_i}$$

$$AVE = \frac{\sum \lambda_i^2}{\sum \lambda_i^2 + \sum \epsilon_i}$$

where
$\lambda_i$   standardized loading for each indicator
$\epsilon_i$   measurement error for each indicator

Discriminant validity is the independence of the dimensions or sub-constructs (Bagozzi and Phillips 1982). To check the discriminant validity, a pair-wise comparison was performed by comparing a model with correlation constrained to one with an unconstrained model. A difference between the $\chi^2$ values (d.f. $= 1$) of the two models that is significant at $p < 0.05$ level would indicate support for the discriminant validity criterion (Joreskog and Sorbom 1986, 1989).

An important aspect of construct validity is the validation of second-order constructs. T coefficient was used to test whether a second-order construct exists accounting for the variations in its sub-constructs. T coefficient is calculated as the ratio of the Chi-square of the first-order model to the Chi-square of the second-order model and a T coefficient of higher than 0.80 indicates the existence of a second-order construct (Doll et al. 1995).

Creating multi-item measures for constructs could adopt a reflective versus formative perspective (Chin 1998; Diamantopoulos 1999; Williams et al. 2003; Patnayakuni et al. 2006). To make a choice between the two views, four criteria are suggested by Jarvis et al. (2003): (1) direction of causality from construct to indicators, (2) interchangeability of indicators, (3) covariation among indicators, and (4) nomological net of construct indicators. Indicators are considered to be reflective when they are manifestations of constructs, are interchangeable and share a common theme, covary with each other. And this nomological net of the indicators are not differing. The opposite conditions would apply in the case of formative indicators.

Constructs, subcomponents, and their indicators can be modeled as either formative (cause) or reflective (effect). Models using formative measures are likely to have difficulties regarding model identification and interpretation (Williams et al. 2003). In this research, a reflective specification is chosen because the subcomponents of each construct are expected to be intercorrelated and covary with each other. SEM program (e.g. LISREL) will be used to validate measures based on reflective indicators. To incorporate both formative and reflective indicators, partial least squares (PLS) approach and SEM can be used.

Finally, a structural analysis using LISREL will be run to test the hypotheses. To assess the fit of the hypothesized model to the data, various fit indices can be used as discussed above. If the model fits the data adequately, the t-values of the gamma and beta coefficients will be evaluated to test the hypotheses. Using one-tailed test, a t-value greater than 2.33 is significant at the level of 0.01; a t-value greater than 1.64 is significant at 0.05.

## 7.2.2 Large-Scale Measurement Results

In the following section, the results of large-scale analysis for each construct will be reported and discussed. The coding for items is shown in Appendices C and D. The instruments that will be used in structural analysis and hypothesis testing after the large-scale study are shown in Appendix F.

**7.2.2.1 IT Resources**

Shown in Table 7.3, the initial fit indices for IT infrastructure flexibility (e.g., RMSEA = 0.090) suggest that improvement could be made in the measures. Examination of the factor loadings and modification indices suggests that IRIF1 should be dropped from the IT infrastructure flexibility scale because of low loading (0.66). IRIF1 (systems are modular) was dropped because respondents may not be clear about what are modular systems and they understand them differently. This explains the low factor loading for IRIF1. After deleting IRIF1, the re-specified one-dimensional model for IT infrastructure flexibility indicates a good fit (CFI = 1.00, NNFI = 0.99, RMSEA = 0.078, normed $\chi^2$ = 2.26). All the factor loadings for the revised constructs are greater than 0.70 and significant at $p < 0.01$ based on t-values. This indicates good convergent validity. The estimate of AVE of 0.70 and the composite reliability of 0.90 exceed the critical value of 0.50 and 0.70 respectively, providing evidence of good reliability.

The same process was followed for IT expertise. The initial fit indices indicate that the improvement can be made in the measures (RMSEA = 0.144, normed $\chi^2$ = 5.35). Examination of the factor loadings and modification indices suggests that IRIE5 should be dropped from the IT expertise scale because of high correlated errors with IRIE4 and IRIE3. The deletion of IRIE5 (our IT staff understand our firm's procedures and policies very well) should have minimal effect on content validity because that portion of the domain is preserved by IRIE4 (understand technologies and business process very well) and IRIE6 (knowledgeable about our business strategies, priorities, and opportunities). The re-specified one-dimensional model for IT expertise indicates a good fit (RMSEA = 0.051, normed $\chi^2$ = 1.54, CFI = 1.00, NNFI = 1.00). All the factor loadings are greater than 0.70 and significant at $p < 0.01$ based on t-values. This indicates good convergent validity. The estimate of AVE of 0.81 and the composite reliability of 0.95 exceed the critical value of 0.50 and 0.70 respectively, providing evidence of good reliability.

Table 7.4 reports the results for 1 pairwise discriminant validity test between the two sub-dimensions of IT resources. The test was run with the correlation between the latent variables fixed at 1.0 and with the correlation between the latent variables unconstrained. The $\chi^2$ difference for 1 degree of freedom is 12.46, significant at $p < 0.01$, and the result strongly supports the case for discriminant validity (Table 7.4).

**7.2.2.2 IOS Appropriation**

Shown in Table 7.5, the initial fit indices for IOS use for integration (e.g., RMSEA = 0.169, normed $\chi^2$ = 6.98) suggest that improvement could be made in the measures. Examination of the factor loadings and modification indices suggests that IAIG1 should be dropped from the IOS use for integration scale because of high correlated errors with IAIG2 and IAIG3. IAIG1 is too general compared to the other four items. The deletion of IAIG1 (The extent of IOS use among supply

**Table 7.3** Confirmatory factor analysis results for IT resources

| Construct | Measurement diagram | Initial measurement model Fit[a] | Modifications Item | Modifications Indication | Modifications Action | Final measurement model fit | Loadings (t statistics) | AVE | $\rho_c$ (Reliability) |
|---|---|---|---|---|---|---|---|---|---|
| IT infrastructure flexibility | | $\chi^2 = 13.55$, df = 5, p = 0.0188, CFI = 0.98, NNFI = 0.97, RMSEA = 0.090, Normed $\chi^2 = 2.71$ | IRIF1 | Low loading (0.66) | Drop IRIF1 | $\chi^2 = 4.52$, df = 2, p = 0.1043, CFI = 1.00, NNFI = 0.99, RMSEA = 0.078, Normed $\chi^2 = 2.26$ | IRIF2: 0.87 (–), IRIF3: 0.88 (16.62), IRIF4: 0.73 (12.41), IRIF5: 0.85 (15.73) | 0.70 | 0.90 |
| IT expertise | | $\chi^2 = 48.19$, df = 9, p = 0.0000, CFI = 0.97, NNFI = 0.95, RMSEA = 0.144, Normed $\chi^2 = 5.35$ | IRIE5 | High correlated error with IRIE4 and IRIE3 | Drop IRIE5 | $\chi^2 = 7.72$, df = 5, p = 0.1725, CFI = 1.00, NNFI = 1.00, RMSEA = 0.051, Normed $\chi^2 = 1.54$ | IRIE1: 0.92 (–), IRIE2: 0.86 (19.37), IRIE3: 0.90 (21.98), IRIE4: 0.88 (20.57), IRIE6: 0.93 (24.06) | 0.81 | 0.95 |

[a] Model fit indices and suggested cut-offs

*CFI* Comparative fit index (>0.90)

*RMSEA* Root mean square error of approximation (<0.08)

NNFI = Non-normed fit index (>0.90)

*Normed* $\chi^2$ Chi-square/degrees of freedom (<3.0)

*AVE* Average variance extracted (>0.50)

$\rho_c$ Reliability (>0.70)

**Table 7.4** Pairwise comparison of $\chi^2$ values for IT resources

| Construct | IRIF | | |
| --- | --- | --- | --- |
| | Free | Fix | Dif. |
| IRIE | 57.60 | 70.06 | **12.46**[a] |

[a] significant at $p < 0.01$; [b] significant at $p < 0.05$; [c] significant at $p < 0.10$

chain partners for integrating business functions across firms) should have minimal effect on content validity because that portion of the domain is preserved by the other four items: IAIG2 (The extent of IOS use among supply chain partners for joint forecasting, planning, and execution), IAIG3 (The extent of IOS use among supply chain partners for order processing, invoicing and settling accounts), IAIG4 (The extent of IOS use among supply chain partners for exchange of shipment and delivery information), and IAIG5 (The extent of IOS use among supply chain partners for managing warehouse stock and inventories). After deleting IAIG1, the re-specified one-dimensional model for IOS use for integration indicates a good fit (RMSEA = 0.000, normed $\chi^2$ = 0.31, CFI = 1.00, NNFI = 1.00). All the factor loadings for the revised constructs are greater than 0.75 and significant at $p < 0.01$ based on t-values. This indicates good convergent validity. The estimate of AVE of 0.68 and the composite reliability of 0.90 exceed the critical value of 0.50 and 0.70 respectively, providing evidence of good reliability.

The same process was followed for IOS use for communication. The initial fit indices indicate that the improvement can be made in the measures (RMSEA = 0.092). Examination of the factor loadings and modification indices suggests that IAIC1 should be dropped from the IOS use for communication scale because of low loading (0.58). IAIC1 (The extent of IOS use among supply chain partners for workflow coordination) was dropped because respondents may think it is an integration issue rather than communication issue. This explains the low factor loading for IAIC1. The re-specified one-dimensional model for IT expertise indicates a good fit (RMSEA = 0.000, normed $\chi^2$ = 0.78, CFI = 1.00, NNFI = 1.00). All the factor loadings are greater than 0.75 and significant at $p < 0.01$ based on t-values. This indicates good convergent validity. The estimate of AVE of 0.64 and the composite reliability of 0.88 exceed the critical value of 0.50 and 0.70 respectively, providing evidence of good reliability.

The same process was followed for IOS use for intelligence. The initial fit indices indicate that the improvement can be made in the measures (RMSEA = 0.198, normed $\chi^2$ = 9.18). Examination of the factor loadings and modification indices suggests that IAIL2 should be dropped from the IOS use for intelligence scale because of high correlated errors with IAIL3 and IAIL4. The deletion of IAIL2 (Our firm and supply chain partners use IOS for storing, searching, and retrieving business information) should have minimal effect on content validity because that portion of the domain is preserved by IAIL4 (Our firm and supply chain partners use IOS for combining information from different sources to uncover trends and patterns). The re-specified one-dimensional model for IOS use for intelligence indicates a good fit (RMSEA = 0.000, normed $\chi^2$ = 0.65,

**Table 7.5** Confirmatory factor analysis results for IOS appropriation

| Construct | Measurement diagram | Initial measurement model fit [a] | Modifications | | | Final measurement model fit | Loadings (t statistics) | AVE | $\rho_c$ (Reliability) |
| --- | --- | --- | --- | --- | --- | --- | --- | --- | --- |
| | | | Item | Indication | Action | | | | |
| IOS use for integration | IOS Use for Integration Flexibility → IAIG1 IAIG2 IAIG3 IAIG4 IAIG5 | $\chi^2 = 34.92$, df = 5, p = 0.0000 CFI = 0.96 NNFI = 0.92 RMSEA = 0.169 Normed $\chi^2 = 6.98$ | IAIG1 | High correlated error with IAIG2 and IAIG3 | Drop IAIG1 | $\chi^2 = 0.61$, df = 2, p = 0.7389 CFI = 1.00 NNFI = 1.00 RMSEA = 0.000 Normed $\chi^2 = 0.31$ | IAIG2: 0.75 (−) IAIG3: 0.89 (13.00) IAIG4: 0.83 (12.12) IAIG5: 0.83 (12.15) | 0.68 | 0.90 |
| IOS use for communication | IOS Use for Communicature Flexibility → IAIC1 IAIC2 IAIC3 IAIC4 IAIC5 | $\chi^2 = 13.88$, df = 5, p = 0.0164 CFI = 0.98 NNFI = 0.96 RMSEA = 0.092 Normed $\chi^2 = 2.78$ | IAIC1 | Low loading (0.58) | Drop IAIC1 | $\chi^2 = 1.56$, df = 2, p = 0.4588 CFI = 1.00 NNFI = 1.00 RMSEA = 0.000 Normed $\chi^2 = 0.78$ | IAIC2: 0.76 (−) IAIC3: 0.87 (12.42) IAIC4: 0.82 (11.87) IAIC5: 0.75 (10.72) | 0.64 | 0.88 |
| IOS for intelligence | IOS for Intelligence → IAIL1 IAIL2 IAIL3 IAIL4 IAIL5 | $\chi^2 = 45.90$, df = 5, p = 0.0000 CFI = 0.96 NNFI = 0.91 RMSEA = 0.198 Normed $\chi^2 = 9.18$ | IAIL2 | High correlated error with IAIL3 and IAIL4 | Drop IAIL2 | $\chi^2 = 1.30$, df = 2, p = 0.5221 CFI = 1.00 NNFI = 1.00 RMSEA = 0.000 Normed $\chi^2 = 0.65$ | IAIL1: 0.82 (−) IAIL3: 0.87 (15.23) IAIL4: 0.89 (15.79) IAIL5: 0.89 (15.73) | 0.75 | 0.92 |

[a] Model fit indices and suggested cut-offs
*CFI* Comparative fit index (>0.90)
*RMSEA* Root mean square error of approximation (<0.08)
*NNFI* Non-normed fit index (>0.90)
*Normed* $\chi^2$ Chi-square/degrees of freedom (<3.0)
*AVE* Average variance extracted (>0.50)
$\rho_c$ Reliability (>0.70)

**Table 7.6** Pairwise comparison of $\chi^2$ values for IOS appropriation IOS appropriation

| Construct | IAIG | | | IAIC | | |
|---|---|---|---|---|---|---|
| | Free | Fix | Dif. | Free | Fix | Dif. |
| IAIC | 60.09 | 66.05 | **5.96**[b] | | | |
| IAIL | 36.82 | 40.33 | **3.51**[c] | 32.08 | 40.23 | **8.15**[a] |

[a] significant at p < 0.01; [b] significant at p < 0.05; [c] significant at p < 0.10

CFI = 1.00, NNFI = 1.00). All the factor loadings are greater than 0.80 and significant at p < 0.01 based on t-values. This indicates good convergent validity. The estimate of AVE of 0.75 and the composite reliability of 0.92 exceed the critical value of 0.50 and 0.70 respectively, providing evidence of good reliability.

Table 7.6 reports the results for 3 pairwise discriminant validity tests between the three sub-dimensions of IOS appropriation. The test was run with the correlation between the latent variables fixed at 1.0 and with the correlation between the latent variables unconstrained. The $\chi^2$ difference between IAIC and IAIL is 8.15, significant at p < 0.01. The $\chi^2$ difference between IAIC and IAIG is 5.96, significant at p < 0.05. The $\chi^2$ difference between IAIG and IAIL is 3.51, significant at p < 0.10. The result supports the case for discriminant validity except the difference between IAIG and IAIL is marginally validated.

### 7.2.2.3 Collaborative Culture

Shown in Table 7.7, the initial fit indices for collectivism (e.g., RMSEA = 0.062, normed $\chi^2$ = 1.79, CFI = 1.00, NNFI = 0.99) suggest that no improvement needs to be made in the measures. The one-dimensional model for collectivism indicates a good fit. All the factor loadings for the constructs are greater than 0.70 and significant at p < 0.01 based on t-values. This indicates good convergent validity. The estimate of AVE of 0.56 and the composite reliability of 0.83 exceed the critical value of 0.50 and 0.70 respectively, providing evidence of good reliability.

The same process was followed for long term orientation. The initial fit indices indicate that no improvement needs to be made in the measures (RMSEA = 0.000, normed $\chi^2$ = 0.37, CFI = 1.00, NNFI = 1.00). The one-dimensional model for long term orientation indicates a good fit. All the factor loadings are greater than 0.70 and significant at p < 0.01 based on t-values. This indicates good convergent validity. The estimate of AVE of 0.68 and the composite reliability of 0.89 exceed the critical value of 0.50 and 0.70 respectively, providing evidence of good reliability.

The same process was followed for power symmetry. The initial fit indices indicate that no improvement needs to be made in the measures (RMSEA = 0.067, normed $\chi^2$ = 1.93, CFI = 1.00, NNFI = 0.99). The one-dimensional model for power symmetry indicates a good fit. All the factor loadings are greater than 0.80 and significant at p < 0.01 based on t-values. This indicates good convergent validity. The estimate of AVE of 0.69 and the composite reliability of

**Table 7.7** Confirmatory factor analysis results for collaborative culture

| Construct | Measurement diagram | Initial measurement model fit[a] | Modifications Item | Indication | Action | Final measurement model fit | Loadings (t statistics) | AVE | $\rho_c$ (Reliability) |
|---|---|---|---|---|---|---|---|---|---|
| Collectivism | | $\chi^2 = 3.58$, df = 2, p = 0.1667, CFI = 1.00, NNFI = 0.99, RMSEA = 0.062, Normed $\chi^2 = 1.79$ | | | No Change | $\chi^2 = 3.58$, df = 2, p = 0.1667, CFI = 1.00, NNFI = 0.99, RMSEA = 0.062, Normed $\chi^2 = 1.79$ | CCCL1: 0.73 (–), CCCL2: 0.72 (9.36), CCCL3: 0.74 (9.55), CCCL4: 0.80 (10.05) | 0.56 | 0.83 |
| Long term orientation | | $\chi^2 = 0.74$, df = 2, p = 0.6904, CFI = 1.00, NNFI = 1.00, RMSEA = 0.000, Normed $\chi^2 = 0.37$ | | | No Change | $\chi^2 = 0.74$, df = 2, p = 0.6904, CFI = 1.00, NNFI = 1.00, RMSEA = 0.000, Normed $\chi^2 = 0.37$ | CCLT1: 0.89 (–), CCLT2: 0.85 (15.79), CCLT3: 0.73 (12.49), CCLT4: 0.81 (14.72) | 0.68 | 0.89 |
| Power symmetry | | $\chi^2 = 3.85$, df = 2, p = 0.1457, CFI = 1.00, NNFI = 0.99, RMSEA = 0.067, Normed $\chi^2 = 1.93$ | | | No Change | $\chi^2 = 3.85$, df = 2, p = 0.1457, CFI = 1.00, NNFI = 0.99, RMSEA = 0.067, Normed $\chi^2 = 1.93$ | CCPS1: 0.83 (–), CCPS2: 0.82 (13.56), CCPS3: 0.84 (13.96), CCPS4: 0.84 (14.06) | 0.69 | 0.90 |

(continued)

**Table 7.7** (continued)

| Construct | Measurement diagram | Initial measurement model fit[a] | Modifications | | | Final measurement model fit | Loadings (t statistics) | AVE | $\rho_c$ (Reliability) |
|---|---|---|---|---|---|---|---|---|---|
| | | | Item | Indication | Action | | | | |
| Uncertainty avoidance | Uncertainty Avoidance → CCUA1, CCUA2, CCUA3, CCUA4 | $\chi^2 = 0.06$, df = 2, $p = 0.9724$ CFI = 1.00 NNFI = 1.00 RMSEA = 0.000 Normed $\chi^2 = 0.03$ | | | No Change | $\chi^2 = 0.06$, df = 2, $p = 0.9724$ CFI = 1.00 NNFI = 1.00 RMSEA = 0.000 Normed $\chi^2 = 0.03$ | CCUA1: 0.67 (–) CCUA2: 0.93 (11.42) CCUA3: 0.85 (10.88) CCUA4: 0.80 (10.30) | 0.67 | 0.89 |

[a] Model fit indices and suggested cut-offs

*CFI* Comparative fit index (>0.90)

*RMSEA* Root mean square error of approximation (<0.08)

*NNFI* non-normed fit index (>0.90)

*Normed* $\chi^2$ Chi-square/degrees of freedom (<3.0)

*AVE* Average variance extracted (>0.50)

$\rho_c$ Reliability (>0.70)

**Table 7.8** Pairwise comparison of $\chi^2$ values for collaborative culture

| Construct | CCCL | | | CCLT | | | CCPS | | |
|---|---|---|---|---|---|---|---|---|---|
| | Free | Fix | Dif. | Free | Fix | Dif. | Free | Fix | Dif. |
| CCLT | 50.44 | 70.91 | 20.47[a] | | | | | | |
| CCPS | 27.32 | 48.01 | 20.69[a] | 14.48 | 294.5 | 280.02[a] | | | |
| CCUA | 50.86 | 85.16 | 34.30[a] | 41.13 | 87.55 | 46.42[a] | 37.95 | 311.75 | 273.80[a] |

[a] significant at $p < 0.01$; [b] significant at $p < 0.05$; [c] significant at $p < 0.10$

0.90 exceed the critical value of 0.50 and 0.70 respectively, providing evidence of good reliability.

The same process was followed for uncertainty avoidance. The initial fit indices indicate that no improvement needs to be made in the measures (RMSEA = 0.000, normed $\chi^2 = 0.03$, CFI = 1.00, NNFI = 1.00). The one-dimensional model for uncertainty avoidance indicates a good fit. All the factor loadings are greater than 0.65 and significant at $p < 0.01$ based on t-values. This indicates good convergent validity. The estimate of AVE of 0.67 and the composite reliability of 0.89 exceed the critical value of 0.50 and 0.70 respectively, providing evidence of good reliability.

Table 7.8 reports the results for six pairwise discriminant validity tests between the 4 sub-dimensions of collaborative culture. The test was run with the correlation between the latent variables fixed at 1.0 and with the correlation between the latent variables unconstrained. The $\chi^2$ differences for one degree of freedom are all significant at $p < 0.01$. The results strongly support the case for discriminant validity.

### 7.2.2.4 Trust

Shown in Table 7.9, the initial fit indices for credibility (e.g., RMSEA = 0.080, normed $\chi^2 = 2.34$, CFI = 0.99, NNFI = 0.98) suggest that no improvement needs to be made in the measures. The one-dimensional model for credibility indicates a good fit. All the factor loadings for the constructs are greater than 0.80 and significant at $p < 0.01$ based on t-values. This indicates good convergent validity. The estimate of AVE of 0.68 and the composite reliability of 0.91 exceed the critical value of 0.50 and 0.70 respectively, providing evidence of good reliability.

The same process was followed for benevolence. The initial fit indices indicate that the improvement can be made in the measures (RMSEA = 0.128, normed $\chi2 = 4.44$). Examination of the factor loadings and modification indices suggests that TRBN4 should be dropped from the benevolence scale because of high correlated errors with TRBN1 and TRBN5. The deletion of TRBN4 (When we share our problems with supply chain partners, we know that they will respond with understanding) should have minimal effect on content validity because that portion of the domain is preserved by TRBN1 (Our supply chain partners have made sacrifices for us in the past) and TRBN5 (We can count on supply chain partners to consider how their actions will affect us). The re-specified one-dimensional model

**Table 7.9** Confirmatory factor analysis results for trust

| Construct | Measurement diagram | Initial measurement model fit[a] | Modifications | | | Final measurement model fit | Loadings (t statistics) | AVE | $\rho_c$ (Reliability) |
|---|---|---|---|---|---|---|---|---|---|
| | | | Item | Indication | Action | | | | |
| Credibility | | $\chi^2 = 11.70$, df = 5, p = 0.0391 CFI = 0.99 NNFI = 0.98 RMSEA = 0.080 Normed $\chi^2 = 2.34$ | | | No Change | $\chi^2 = 11.70$, df = 5, p = 0.0391 CFI = 0.99 NNFI = 0.98 RMSEA = 0.080 Normed $\chi^2 = 2.34$ | TRCR1: 0.81 (–) TRCR2: 0.86 (14.31) TRCR3: 0.81 (13.29) TRCR4: 0.83 (13.60) TRCR5: 0.82 (13.36) | 0.68 | 0.91 |
| Benevolence | | $\chi^2 = 22.22$, df = 5, p = 0.005 CFI = 0.98 NNFI = 0.96 RMSEA = 0.128 Normed $\chi^2 = 4.44$ | TRBN4 | High correlated error with TRBN1 and TRBN5 | Drop TRBN4 | $\chi^2 = 2.45$, df = 2, p = 0.2938 CFI = 1.00 NNFI = 1.00 RMSEA = 0.033 Normed $\chi^2 = 1.23$ | TRBN1: 0.85 (–) TRBN2: 0.80 (13.98) TRBN3: 0.92 (17.62) TRBN5: 0.90 (17.24) | 0.75 | 0.92 |

[a] Model fit indices and suggested cut-offs
*CFI* Comparative fit index (>0.90)
*RMSEA* Root mean square error of approximation (<0.08)
*NNFI* Non-normed fit index (>0.90)
*Normed* $\chi^2$ Chi-square/degrees of freedom (<3.0)
*AVE* Average variance extracted (>0.50)
$\rho_c$ Reliability (>0.70)

**Table 7.10** Pairwise comparison of $\chi^2$ Values for Trust

| Construct | TRCR | | |
| | Free | Fix | Dif. |
| --- | --- | --- | --- |
| TRBN | 58.10 | 67.29 | **9.19**[a] |

[a] significant at $p < 0.01$; [b] significant at $p < 0.05$; [c] significant at $p < 0.10$

for benevolence indicates a good fit (RMSEA = 0.033, normed $\chi^2$ = 1.23, CFI = 1.00, NNFI = 1.00). All the factor loadings are greater than 0.80 and significant at $p < 0.01$ based on t-values. This indicates good convergent validity. The estimate of AVE of 0.75 and the composite reliability of 0.92 exceed the critical value of 0.50 and 0.70 respectively, providing evidence of good reliability.

Table 7.10 reports the results for 1 pairwise discriminant validity test between the two sub-dimensions of IT resources. The test was run with the correlation between the latent variables fixed at 1.0 and with the correlation between the latent variables unconstrained. The $\chi^2$ difference for 1 degree of freedom is 9.19, significant at $p < 0.01$, and the result strongly supports the case for discriminant validity.

### 7.2.2.5 Supply Chain Collaboration

Shown in Table 7.11, the initial fit indices for information sharing (e.g., RMSEA = 0.054, normed $\chi^2$ = 1.60, CFI = 0.99, NNFI = 0.99) suggest that no improvement needs to be made in the measures. But a closer examination of the factor loadings suggests that SCIS1 should be dropped from the information sharing scale because of low loading (0.56). SCIS1 (Our firm and supply chain partners exchange relevant information) was dropped because respondents may understand relevant information differently. This explains the low factor loading for SCIS1. After deleting SCIS1, the re-specified one-dimensional model for information sharing indicates a good fit (RMSEA = 0.000, normed $\chi^2$ = 0.77, CFI = 1.00, NNFI = 1.00). All the factor loadings for the revised constructs are greater than 0.75 and significant at $p < 0.01$ based on t-values. This indicates good convergent validity. The estimate of AVE of 0.60 and the composite reliability of 0.86 exceed the critical value of 0.50 and 0.70 respectively, providing evidence of good reliability.

The same process was followed for goal congruence. The initial fit indices indicate that improvement can be made in the measures (RMSEA = 0.108, normed $\chi^2$ = 3.43). Examination of the factor loadings and modification indices suggests that SCGC5 should be dropped from the goal congruence scale because of high correlated errors with SCGC1. The deletion of SCGC5 (Our firm and supply chain partners jointly layout collaboration implementation plans to achieve the goals of the supply chain) should have minimal effect on content validity because that portion of the domain is preserved by SCGC1 (Our firm and supply chain partners have agreement on the goals of the supply chain). The re-specified

**Table 7.11** Confirmatory factor analysis results for supply chain collaboration

| Construct | Measurement diagram | Initial measurement model fit[a] | Modifications Item | Indication | Action | Final measurement model fit | Loadings (t statistics) | AVE | $\rho_c$ (Reliability) |
|---|---|---|---|---|---|---|---|---|---|
| Information sharing | Information Sharing (SCIS1, SCIS2, SCIS3, SCIS4, SCIS5) | $\chi^2 = 8$, df = 5, p = 0.1564, CFI = 0.99, NNFI = 0.99, RMSEA = 0.054, Normed $\chi^2 = 1.60$ | SCIS1 | Low loading (0.56) | Drop SCIS1 | $\chi^2 = 1.54$, df = 2, p = 0.4628, CFI = 1.00, NNFI = 1.00, RMSEA = 0.000, Normed $\chi^2 = 0.77$ | SCIS2: 0.87 (–), SCIS3: 0.78 (12.41), SCIS4: 0.76 (12.06), SCIS5: 0.68 (10.53) | 0.60 | 0.86 |
| Goal congruence | Goal Congruence (SCGC1, SCGC2, SCGC3, SCGC4, SCGC5) | $\chi^2 = 17.16$, df = 5, p = 0.0042, CFI = 0.98, NNFI = 0.96, RMSEA = 0.108, Normed $\chi^2 = 3.43$ | SCGC5 | High correlated error with SCGC1 | Drop SCGC5 | $\chi^2 = 1.52$, df = 2, p = 0.4671, CFI = 1.00, NNFI = 1.00, RMSEA = 0.000, Normed $\chi^2 = 0.76$ | SCGC1: 0.85 (–), SCGC2: 0.76 (11.82), SCGC3: 0.77 (12.01), SCGC4: 0.78 (12.30) | 0.63 | 0.87 |
| Decision synchronization | Decision Synchronization (SCDS1, SCDS2, SCDS3, SCDS4, SCDS5) | $\chi^2 = 18.33$, df = 5, p = 0.0026, CFI = 0.97, NNFI = 0.94, RMSEA = 0.113, Normed $\chi^2 = 3.67$ | SCDS5 | High correlated error with SCDS3 | Drop SCDS5 | $\chi^2 = 1.32$, df = 2, p = 0.5156, CFI = 1.00, NNFI = 1.00, RMSEA = 0.000, Normed $\chi^2 = 0.66$ | SCDS1: 0.72 (–), SCDS2: 0.79 (9.66), SCDS3: 0.71 (8.58), SCDS4: 0.75 (9.38) | 0.55 | 0.83 |

(continued)

**Table 7.11** (continued)

| Construct | Measurement diagram | Initial measurement model fit[a] | Modifications | | | Final measurement model fit | Loadings (t statistics) | AVE | $\rho_c$ (Reliability) |
|---|---|---|---|---|---|---|---|---|---|
| | | | Item | Indication | Action | | | | |
| Incentive alignment | Incentive Alignment → SCIA1, SCIA2, SCIA3, SCIA4, SCIA5 | $\chi^2 = 17.52$, df = 5, p = 0.0036, CFI = 0.97, NNFI = 0.95, RMSEA = 0.109, Normed $\chi^2 = 3.50$ | SCIA3 | High correlated error with SCIA4 | Drop SCIA3 | $\chi^2 = 3.68$, df = 2, p = 0.1587, CFI = 0.99, NNFI = 0.98, RMSEA = 0.063, Normed $\chi^2 = 1.84$ | SCIA1: 0.75 (–), SCIA2: 0.84 (10.84), SCIA4: 0.71 (9.52), SCIA5: 0.73 (9.80) | 0.58 | 0.84 |
| Resource sharing | Resource Sharing → SCRS1, SCRS2, SCRS3, SCRS4, SCRS5 | $\chi^2 = 7.86$, df = 5, p = 0.1643, CFI = 0.99, NNFI = 0.99, RMSEA = 0.052, Normed $\chi^2 = 1.57$ | SCRS2 | Low loading (0.56) | Drop SCRS2 | $\chi^2 = 2.45$, df = 2, p = 0.2932, CFI = 1.00, NNFI = 1.00, RMSEA = 0.033, Normed $\chi^2 = 1.23$ | SCRS1: 0.70 (–), SCRS3: 0.80 (10.27), SCRS4: 0.79 (10.13), SCRS5: 0.83 (10.53) | 0.61 | 0.86 |
| Collaborative communication | Collaborative Communication → SCCM1, SCCM2, SCCM3, SCCM4, SCCM5 | $\chi^2 = 5.76$, df = 5, p = 0.3299, CFI = 1.00, NNFI = 1.00, RMSEA = 0.027, Normed $\chi^2 = 1.15$ | | | No Change | $\chi^2 = 5.76$, df = 5, p = 0.3299, CFI = 1.00, NNFI = 1.00, RMSEA = 0.027, Normed $\chi^2 = 1.15$ | SCCM1: 0.89 (–), SCCM2: 0.87 (16.94), SCCM3: 0.72 (12.41), SCCM4: 0.78 (14.16), SCCM5: 0.71 (12.04) | 0.60 | 0.85 |

(continued)

**Table 7.11** (continued)

| Construct | Measurement diagram | Initial measurement model fit[a] | Modifications | | Action | Final measurement model fit | Loadings (t statistics) | AVE | $\rho_c$ (Reliability) |
|---|---|---|---|---|---|---|---|---|---|
| | | | Item | Indication | | | | | |
| Joint knowledge creation | Joint Knowledge Creation → SCKC1, SCKC2, SCKC3, SCKC4, SCKC5 | $\chi^2 = 8.38$, df $= 5$, p $= 0.1365$<br>CFI $= 1.00$<br>NNFI $= 0.99$<br>RMSEA $= 0.057$<br>Normed $\chi^2 = 1.68$ | | | No Change | $\chi^2 = 8.38$, df $= 5$, p $= 0.1365$<br>CFI $= 1.00$<br>NNFI $= 0.99$<br>RMSEA $= 0.057$<br>Normed $\chi^2 = 1.68$ | SCKC1: 0.89<br>(–)<br>SCKC2: 0.88<br>(17.66)<br>SCKC3: 0.81<br>(15.39)<br>SCKC4: 0.75<br>(13.36)<br>SCKC5: 0.76<br>(13.72) | 0.58 | 0.88 |

[a] Model fit indices and suggested cut-offs
*CFI* Comparative fit index (>0.90)
*RMSEA* Root mean square error of approximation (<0.08)
*NNFI* Non-normed fit index (>0.90)
*Normed $\chi^2$* = Chi-square/degrees of freedom (<3.0)
*AVE* Average variance extracted (>0.50)
*$\rho_c$* Reliability (>0.70)

one-dimensional model for the construct indicates a good fit (RMSEA = 0.000, normed $\chi^2$ = 0.76, CFI = 1.00, NNFI = 1.00). All the factor loadings are greater than 0.75 and significant at p < 0.01 based on t-values. This indicates good convergent validity. The estimate of AVE of 0.63 and the composite reliability of 0.87 exceed the critical value of 0.50 and 0.70 respectively, providing evidence of good reliability.

The same process was followed for decision synchronization. The initial fit indices indicate that improvement can be made in the measures (RMSEA = 0.113, normed $\chi^2$ = 3.67). Examination of the factor loadings and modification indices suggests that SCDS5 should be dropped from the decision synchronization scale because of high correlated errors with SCDS3. The deletion of SCDS5 (Our firm and supply chain partners jointly work out solutions) should have minimal effect on content validity because that portion of the domain is preserved by SCGC1 (Our firm and supply chain partners jointly plan on promotional events), SCGC2 (Our firm and supply chain partners jointly develop demand forecasts), SCGC3 (Our firm and supply chain partners jointly manage inventory), and SCGC4 (Our firm and supply chain partners jointly plan on product assortment). The re-specified one-dimensional model for the construct indicates a good fit (RMSEA = 0.000, normed $\chi^2$ = 0.66, CFI = 1.00, NNFI = 1.00). All the factor loadings are greater than 0.70 and significant at p < 0.01 based on t-values. This indicates good convergent validity. The estimate of AVE of 0.55 and the composite reliability of 0.83 exceed the critical value of 0.50 and 0.70 respectively, providing evidence of good reliability.

The same process was followed for incentive alignment. The initial fit indices indicate improvement can be made in the measures (RMSEA = 0.109, normed $\chi^2$ = 3.50). Examination of the factor loadings and modification indices suggests that SCIA3 should be dropped from the incentive alignment scale because of high correlated errors with SCIA4. The deletion of SCIA3 (Our firm and supply chain partners co-develop systems to evaluate and publicize each other's performance) was dropped because it is more related to collaborative performance than incentive alignment. The re-specified one-dimensional model for incentive alignment indicates a good fit (RMSEA = 0.063, normed $\chi^2$ = 1.84, CFI = 0.99, NNFI = 0.98). All the factor loadings are greater than 0.70 and significant at p < 0.01 based on t-values. This indicates good convergent validity. The estimate of AVE of 0.58 and the composite reliability of 0.84 exceed the critical value of 0.50 and 0.70 respectively, providing evidence of good reliability.

The same process was followed for resource sharing. The initial fit indices for resource sharing (e.g., RMSEA = 0.052, normed $\chi^2$ = 1.57, CFI = 0.99, NNFI = 0.99) suggest that no improvement is needed in the measures. A closer examination of the factor loadings suggests SCRS2 be dropped from the resource sharing scale because of low loading (0.56). The deletion of SCRS2 (Our firm and supply chain partners dedicate personnel to manage the collaborative processes) should have minimal effect on content validity because that portion of the domain is preserved by SCRS1 (Our firm and supply chain partners use cross-organizational teams frequently for process design and improvement), SCRS3 (Our firm and supply chain partners share technical supports), and SCRS5 (Our firm and supply chain partners pool financial and non-financial resources). After SCRS2 was deleted, the re-specified one-dimensional model for the construct indicates a good fit (RMSEA = 0.033, normed $\chi^2$ = 1.23, CFI = 1.00, NNFI = 1.00). All the factor loadings are greater than 0.70 and significant at $p < 0.01$ based on t-values. This indicates good convergent validity. The estimate of AVE of 0.61 and the composite reliability of 0.86 exceed the critical value of 0.50 and 0.70 respectively, providing evidence of good reliability.

The same process was followed for collaborative communication. The initial fit indices for collaborative communication (e.g., RMSEA = 0.027, normed $\chi^2$ = 1.15, CFI = 1.00, NNFI = 1.00) suggest that no improvement needs to be made in the measures. The one-dimensional model for the construct indicates a good fit. All the factor loadings for the construct are greater than 0.70 and significant at $p < 0.01$ based on t-values. This indicates good convergent validity. The estimate of AVE of 0.60 and the composite reliability of 0.85 exceed the critical value of 0.50 and 0.70 respectively, providing evidence of good reliability.

The same process was followed for joint knowledge creation. The initial fit indices for joint knowledge creation (e.g., RMSEA = 0.057, normed $\chi^2$ = 1.68, CFI = 1.00, NNFI = 0.99) suggest that no improvement is needed in the measures. The one-dimensional model for the construct indicates a good fit. All the factor loadings for the construct are greater than 0.75 and significant at $p < 0.01$ based on t-values. This indicates good convergent validity. The estimate of AVE of 0.58 and the composite reliability of 0.88 exceed the critical value of 0.50 and 0.70 respectively, providing evidence of good reliability.

Table 7.12 reports the results for 21 pairwise discriminant validity tests between the 7 sub-dimensions of supply chain collaboration. The test was run with the correlation between the latent variables fixed at 1.0 and with the correlation between the latent variables unconstrained. The $\chi^2$ differences are all significant at $p < 0.05$ with 13 out of 21 pairwise comparisons are significant at $p < 0.01$. The results support the case for discriminant validity.

### 7.2.2.6 Collaborative Advantage

Shown in Table 7.13, the initial fit indices for process efficiency (e.g., RMSEA = 0.065, normed $\chi^2$ = 1.89, CFI = 1.00, NNFI = 0.99) suggest that no improvement needs to be made in the measures. The one-dimensional model for process efficiency indicates a good fit. All the factor loadings for the constructs are greater than 0.75 and significant at $p < 0.01$ based on t-values. This indicates good convergent validity. The estimate of AVE of 0.66 and the composite reliability of 0.89 exceed the critical value of 0.50 and 0.70 respectively, providing evidence of good reliability.

The same process was followed for offering flexibility. The initial fit indices indicate that no improvement needs to be made in the measures (RMSEA = 0.000, normed $\chi^2$ = 0.52, CFI = 1.00, NNFI = 1.00). The one-dimensional model for

**Table 7.12** Pairwise comparison of $\chi^2$ values for supply chain collaboration

| Construct | SCIS | | | SCGC | | | SCDS | | | SCIA | | | SCRS | | | SCCM | | |
|---|---|---|---|---|---|---|---|---|---|---|---|---|---|---|---|---|---|---|
| | Free | Fix | Dif. | Free | Fix | Dif. | Free | Fix | Dif. | Free | Fix | Dif. | Free | Fix | Dif. | Free | Fix | Dif. |
| SCGC | 37.65 | 41.89 | **4.24**[b] | | | | | | | | | | | | | | | |
| SCDS | 65.47 | 74.11 | **8.64**[a] | 20.29 | 26.71 | **6.42**[b] | | | | | | | | | | | | |
| SCIA | 23.65 | 30.80 | **7.15**[a] | 64.29 | 76.02 | **11.73**[a] | 59.78 | 66.64 | **6.86**[a] | | | | | | | | | |
| SCRS | 24.94 | 32.41 | **7.47**[a] | 49.10 | 61.59 | **12.49**[a] | 40.85 | 51.41 | **10.56**[a] | 43.73 | 53.47 | **9.74**[a] | | | | | | |
| SCCM | 47.15 | 53.20 | **6.05**[b] | 48.91 | 53.95 | **5.04**[b] | 78.38 | 87.97 | **9.59**[a] | 68.25 | 75.46 | **7.21**[a] | 43.51 | 51.80 | **8.29**[a] | | | |
| SCKC | 51.91 | 56.78 | **4.87**[b] | 49.36 | 55.86 | **6.50**[b] | 43.85 | 50.74 | **6.89**[a] | 31.19 | 36.29 | **5.10**[b] | 39.49 | 49.90 | **10.41**[a] | 51.43 | 57.09 | **5.66**[b] |

[a] significant at $p < 0.01$; [b] significant at $p < 0.05$; [c] significant at $p < 0.10$

**Table 7.13** Confirmatory factor analysis results for collaborative advantage

| Construct | Measurement diagram | Initial measurement model fit[a] | Modifications Item | Indication | Action | Final measurement model fit | Loadings (t statistics) | AVE | $\rho_c$ (Reliability) |
|---|---|---|---|---|---|---|---|---|---|
| Process efficiency | Process Efficiency (CAPE1, CAPE2, CAPE3, CAPE4) | $\chi^2 = 3.78$, df = 2, p = 0.1509<br>CFI = 1.00<br>NNFI = 0.99<br>RMSEA = 0.065<br>Normed $\chi^2 = 1.89$ | | | No Change | $\chi^2 = 3.78$, df = 2, p = 0.1509<br>CFI = 1.00<br>NNFI = 0.99<br>RMSEA = 0.065<br>Normed $\chi^2 = 1.89$ | CAPE1: 0.85 (–)<br>CAPE2: 0.79 (13.06)<br>CAPE3: 0.83 (13.88)<br>CAPE4: 0.79 (12.88) | 0.66 | 0.89 |
| Offering flexibility | Offering Flexibility (CAOF1, CAOF2, CAOF3, CAOF4) | $\chi^2 = 1.03$, df = 2, p = 0.5978<br>CFI = 1.00<br>NNFI = 1.00<br>RMSEA = 0.000<br>Normed $\chi^2 = 0.52$ | | | No Change | $\chi^2 = 1.03$, df = 2, p = 0.5978<br>CFI = 1.00<br>NNFI = 1.00<br>RMSEA = 0.000<br>Normed $\chi^2 = 0.52$ | CAOF1: 0.92 (–)<br>CAOF2: 0.90 (20.56)<br>CAOF3: 0.87 (19.31)<br>CAOF4: 0.81 (16.50) | 0.77 | 0.93 |
| Business synergy | Business Synergy (CABS1, CABS2, CABS3, CABS4) | $\chi^2 = 0.59$, df = 2, p = 0.7446<br>CFI = 1.00<br>NNFI = 1.00<br>RMSEA = 0.000<br>Normed $\chi^2 = 0.30$ | | | No Change | $\chi^2 = 0.59$, df = 2, p = 0.7446<br>CFI = 1.00<br>NNFI = 1.00<br>RMSEA = 0.000<br>Normed $\chi^2 = 0.30$ | CABS1: 0.85 (–)<br>CABS2: 0.86 (15.57)<br>CABS3: 0.85 (15.14)<br>CABS4: 0.87 (15.79) | 0.74 | 0.92 |
| Quality | Quality (CAQL1, CAQL2, CAQL3, CAQL4) | $\chi^2 = 1.76$, df = 2, p = 0.4155<br>CFI = 1.00<br>NNFI = 1.00<br>RMSEA = 0.000<br>Normed $\chi^2 = 0.88$ | | | No Change | $\chi^2 = 1.76$, df = 2, p = 0.4155<br>CFI = 1.00<br>NNFI = 1.00<br>RMSEA = 0.000<br>Normed $\chi^2 = 0.88$ | CAQL1: 0.92 (–)<br>CAQL2: 0.90 (19.89)<br>CAQL3: 0.88 (19.31)<br>CAQL4: 0.71 (12.75) | 0.73 | 0.92 |

(continued)

**Table 7.13** (continued)

| Construct | Measurement diagram | Initial measurement model fit[a] | Modifications | | | Final measurement model fit | Loadings (t statistics) | AVE | $\rho_c$ (Reliability) |
|---|---|---|---|---|---|---|---|---|---|
| | | | Item | Indication | Action | | | | |
| Innovation | Innovation → CAIN1, CAIN2, CAIN3, CAIN4 | $\chi^2 = 0.37$, df = 2, p = 08310<br>CFI = 1.00<br>NNFI = 1.00<br>RMSEA = 0.000<br>Normed $\chi^2 = 0.19$ | | | No Change | $\chi^2 = 0.37$, df = 2, p = 08310<br>CFI = 1.00<br>NNFI = 1.00<br>RMSEA = 0.000<br>Normed $\chi^2 = 0.19$ | CAIN1: 0.88 (–)<br>CAIN2: 0.87 (16.21)<br>CAIN3: 0.81 (14.58)<br>CAIN4: 0.82 (14.93) | 0.71 | 0.91 |

[a] Model fit indices and suggested cut-offs
*CFI* Comparative fit index (>0.90)
*RMSEA* Root mean square error of approximation (<0.08)
*NNFI* Non-normed fit index (>0.90)
*Normed $\chi^2$* Chi-square/degrees of freedom (<3.0)
*AVE* Average variance extracted (>0.50)
$\rho_c$ Reliability (>0.70)

offering flexibility indicates a good fit. All the factor loadings are greater than 0.80 and significant at p < 0.01 based on t-values. This indicates good convergent validity. The estimate of AVE of 0.77 and the composite reliability of 0.93 exceed the critical value of 0.50 and 0.70 respectively, providing evidence of good reliability.

The same process was followed for business synergy. The initial fit indices indicate that no improvement needs to be made in the measures (RMSEA = 0.000, normed $\chi^2$ = 0.30, CFI = 1.00, NNFI = 1.00). The one-dimensional model for business synergy indicates a good fit. All the factor loadings are greater than 0.85 and significant at p < 0.01 based on t-values. This indicates good convergent validity. The estimate of AVE of 0.74 and the composite reliability of 0.92 exceed the critical value of 0.50 and 0.70 respectively, providing evidence of good reliability.

The same process was followed for quality. The initial fit indices indicate that no improvement needs to be made in the measures (RMSEA = 0.000, normed $\chi^2$ = 0.88, CFI = 1.00, NNFI = 1.00). The one-dimensional model for quality indicates a good fit. All the factor loadings are greater than 0.65 and significant at p < 0.01 based on t-values. This indicates good convergent validity. The estimate of AVE of 0.73 and the composite reliability of 0.92 exceed the critical value of 0.50 and 0.70 respectively, providing evidence of good reliability.

The same process was followed for innovation. The initial fit indices indicate that no improvement needs to be made in the measures (RMSEA = 0.000, normed $\chi^2$ = 0.19, CFI = 1.00, NNFI = 1.00). The one-dimensional model for innovation indicates a good fit. All the factor loadings are greater than 0.80 and significant at p < 0.01 based on t-values. This indicates good convergent validity. The estimate of AVE of 0.71 and the composite reliability of 0.91 exceed the critical value of 0.50 and 0.70 respectively, providing evidence of good reliability.

Table 7.14 reports the results for 10 pairwise discriminant validity tests between the 5 sub-dimensions of collaborative advantage. The test was run with the correlation between the latent variables fixed at 1.0 and with the correlation between the latent variables unconstrained. The $\chi^2$ differences are all significant at p < 0.01. The results strongly support the case for discriminant validity.

### 7.2.2.7  Firm Performance

Shown in Table 7.15, the initial fit indices for firm performance (e.g., RMSEA = 0.194, normed $\chi^2$ = 8.86) suggest that improvements be made in the measures. Examination of the factor loadings and modification indices suggests that FP2 should be dropped from the firm performance scale because of high correlated errors with FP3, FP4, and FP6. After deleting FP2, the model was re-run with the remaining 6 items, the results indicate FP7 should be dropped because of high correlated errors with FP1 and FP4. After dropping FP7, the model was re-run with the remaining 5 indicators. The results indicate that FP1 should be dropped because of high correlated errors with FP4 and FP5. Based on the literature, there is no significant relationship between return on investment or profitability and

**Table 7.14** Pairwise comparison of $\chi^2$ values for collaborative advantage

| Construct | CAPE | | | CAOF | | | CABS | | | CAQL | | |
|---|---|---|---|---|---|---|---|---|---|---|---|---|
| | Free | Fix | Dif. | Free | Fix | Dif. | Free | Fix | Dif. | Free | Fix | Dif. |
| CAOF | 20.57 | 31.81 | **11.24**[a] | | | | | | | | | |
| CABS | 23.64 | 31.96 | **8.32**[a] | 44.61 | 93.83 | **49.22**[a] | | | | | | |
| CAQL | 46.25 | 63.29 | **17.04**[a] | 32.03 | 46.81 | **14.78**[a] | 43.41 | 65.81 | **22.40**[a] | | | |
| CAIN | 17.62 | 28.52 | **10.90**[a] | 35.12 | 45.65 | **10.53**[a] | 47.80 | 57.23 | **9.43**[a] | 93.50 | 107.68 | **14.18**[a] |

[a] significant at $p < 0.01$; [b] significant at $p < 0.05$; [c] significant at $p < 0.10$

**Table 7.15** Confirmatory factor analysis results for firm performance

| Construct | Measurement diagram | Initial measurement model fit[a] | Modifications Item | Indication | Action | Final measurement model fit | Loadings (t statistics) | AVE | $\rho_c$ (Reliability) |
|---|---|---|---|---|---|---|---|---|---|
| Firm performance | | $\chi^2 = 124.07$, df = 14, p = 0.0000 CFI = 0.90 NNFI = 0.85 RMSEA = 0.194 Normed $\chi^2 = 8.86$ | FP2 | High correlated error with FP1, FP3, FP4, FP6 | Drop FP2 | $\chi^2 = 0.40$, df = 2, p = 0.8198 CFI = 0.99 NNFI = 0.98 RMSEA = 0.000 Normed $\chi^2 = 0.20$ | FP3: 0.80 (–) FP4: 0.95 (16.16) FP5: 0.87 (14.55) FP6: 0.81 (13.26) | 0.74 | 0.92 |
| | | | FP7 | High correlated error with FP1, FP4 | Drop FP7 | | | | |
| | | | FP1 | High correlated error with FP4, FP5 | Drop FP1 | | | | |

[a] Model fit indices and suggested cut-offs

*CFI* Comparative fit index (>0.90)

*RMSEA* Root mean square error of approximation (<0.08)

*NNFI* Non-normed fit index (>0.90)

*Normed* $\chi^2$ Chi-square/degrees of freedom (<3.0)

*AVE* Average variance extracted (>0.50)

$\rho_c$ Reliability (>0.70)

market share (Anterasian et al. 1996; Vishwanath and Mark 1997). So the deletion of FP2 (growth of market share), FP7 (overall competitive position), and FP1 (market share) should have minimal effect on content validity because the remaining four items capture the return on investment, sales, and profitability. The re-specified one-dimensional model for firm performance indicates a good fit (RMSEA = 0.000, normed $\chi^2$ = 0.20, CFI = 1.00, NNFI = 1.00). All the factor loadings are greater than 0.80 and significant at p < 0.01 based on t-values. This indicates good convergent validity. The estimate of AVE of 0.74 and the composite reliability of 0.92 exceed the critical value of 0.50 and 0.70 respectively, providing evidence of good reliability.

### 7.2.2.8 Validation of Second-Order Constructs

The second-order model explains the co-variations among first-order factors in a more parsimonious way. However, the variations shared by the first-order factors cannot be totally explained by the single second-order factor, and thus the fit indices of the higher-order model can never be better than the corresponding first-order model (Segars and Grover 1998). The first-order model provides a target fit for higher-order models. The efficacy of second-order models can be assessed by examining the target (T) coefficient (where T = first-order $\chi^2$/second-order $\chi^2$) (Marsh and Hocevar 1985). The T coefficient 0.80 to 1.00 indicates the existence of a second-order construct.

Table 7.16 shows the calculated target coefficient between the first-order model and the second-order model for each construct discussed here. Because IT resources and trust each has only two sub-constructs, the fit indices for their first order and second order models are the same and their T-coefficients are 1.0. All the other T-coefficients in Table 4.15 are between 0.9 and 1.0, suggesting that the second-order models should be accepted as more accurate representation of model structure over the corresponding first-order models because they represent more parsimonious explanation of observed covariance. The results support the second-order constructs proposed in theory development sections.

## 7.3 Structural Analysis and Hypothesis Testing

To assess the suggested relationships shown in Fig. 2.1, a structural LISREL model was built. First, the aggregate score of the items factorially loaded for each sub-construct was computed. Second, the sub-construct's aggregate score was used as indicators for the corresponding construct. Third, the structural relationships between constructs were specified as shown in Fig. 7.1. IT resources, collaborative culture, and trust are exogenous variables. The endogenous variables include IOS appropriation, supply chain collaboration, collaborative advantage, and firm performance. Endogenous latent variables are affected by the exogenous variables in the model directly or indirectly.

**Table 7.16** Fit indices for first and second order model

| Construct | Model | $\chi^2$ (df) | Normed $\chi^2$ | CFI | NNFI | RMSEA | T Coefficient (%) |
|---|---|---|---|---|---|---|---|
| IT resources | First-order | 57.60 (26) | 0.29 | 0.98 | 0.97 | 0.076 | 100 |
| | Second-order | 57.60 (26) | 0.29 | 0.98 | 0.97 | 0.076 | |
| IOS appropriation | First-order | 114.87 (51) | 2.25 | 0.96 | 0.95 | 0.077 | 98.72 |
| | Second-order | 117.42 (52) | 2.26 | 0.96 | 095 | 0.078 | |
| Collaborative culture | First-order | 211.08 (98) | 2.15 | 0.95 | 0.94 | 0.074 | 96.10 |
| | Second-order | 222.69 (100) | 2.23 | 0.94 | 0.93 | 0.077 | |
| Trust | First-order | 58.10 (26) | 2.23 | 0.98 | 0.97 | 0.077 | 100 |
| | Second-order | 58.10 (26) | 2.23 | 0.98 | 0.97 | 0.077 | |
| Supply chain collaboration | First-order | 836.62 (384) | 2.18 | 0.89 | 0.87 | 0.075 | 97.40 |
| | Second-order | 887.42 (398) | 2.23 | 0.88 | 0.86 | 0.077 | |
| Collaborative advantage | First-order | 344.11 (160) | 2.15 | 0.94 | 0.93 | 0.074 | 92.50 |
| | Second-order | 384.38 (165) | 2.33 | 0.93 | 0.92 | 0.080 | |

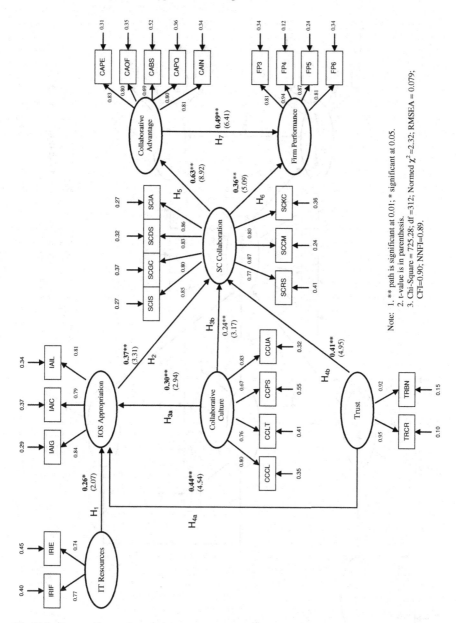

**Fig. 7.1** Structural equation model results for the proposed model

To further validate the proposed model, an alternative model was also tested. It is argued that trust may have a direct impact on collaborative culture (Litwinenko and Copper 1994) and may have an indirect impact on IOS appropriation. The alternative model is specified in Fig. 7.2.

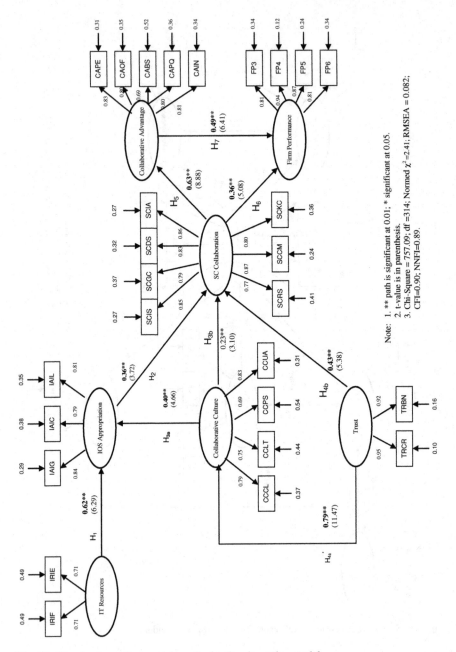

**Fig. 7.2** Structural equation model results for the alternative model

**Table 7.17** Structural modeling results

| Hypotheses | Relationship | Path coefficients | t-Value | Supported |
|---|---|---|---|---|
| $H_1$ | IR → IA | 0.26 | 2.07 | Yes |
| $H_2$ | IA → SC | 0.37 | 3.26 | Yes |
| $H_{3a}$ | CC → IA | 0.30 | 2.94 | Yes |
| $H_{3b}$ | CC → SC | 0.24 | 3.17 | Yes |
| $H_{4a}$ | TR → IA | 0.44 | 4.54 | Yes |
| $H_{4b}$ | TR → SC | 0.41 | 4.95 | Yes |
| $H_5$ | SC → CA | 0.63 | 8.92 | Yes |
| $H_6$ | SC → FP | 0.36 | 5.09 | Yes |
| $H_7$ | CA → FP | 0.49 | 6.41 | Yes |

## 7.3.1 Results of Structural Analysis and Hypotheses Testing

The LISREL results (Fig. 7.1) indicate that the proposed model has a good fit with Chi-square = 725.28 and d.f. = 312, resulting a normed Chi-square of 2.32. RMSEA is 0.079, and CFI and NNFI are 0.90 and 0.89 respectively.

The LISREL results (Fig. 7.2) of the alternative model has a Chi-square = 757.09 and d.f. = 314, resulting a normed Chi-square of 2.41. RMSEA is 0.082, and CFI and NNFI are 0.90 and 0.89 respectively. Based on the values of normed Chi-square and RMSEA, the proposed model performs better.

Between these two models tested, the data better supports the proposed model in Fig. 7.1. The findings for the proposed structural model are summarized in Table 7.17. Eight out of nine hypothesized relationships are strongly supported with the significant, direct positive effects at the 0.01 level. These hypotheses include $H_2$ (IOS appropriation to supply chain collaboration), $H_{3a}$ (collaborative culture to IOS appropriation), $H_{3b}$ (collaborative culture to supply chain collaboration), $H_{4a}$ (trust to IOS appropriation), $H_{4b}$ (trust to supply chain collaboration), $H_5$ (supply chain collaboration to collaborative advantage), $H_6$ (supply chain collaboration to firm performance) and $H_7$ (collaborative advantage to firm performance). The path coefficients and t-values for these hypotheses are respectively 0.37(3.26), 0.30(2.94), 0.24(3.17), 0.44(4.54), 0.41(4.95), 0.63(8.92), 0.36(5.09), and 0.49(6.41). $H_1$ (IT resources to IOS appropriation) is supported with the significant, direct positive effect (path coefficient = 0.26, t-value = 2.07) at the 0.05 level.

According to Joreskog and Sorbom (1986), it is helpful to study relationships by breaking total effects into direct and indirect. To examine the total and component effects, all the coefficients are calculated, shown in Table 7.18. The hypotheses with direct effects are already discussed. A closer look at the hypotheses with both direct and indirect effects in Table 7.18 is needed.

It was hypothesized that collaborative culture has a significant positive relationship with supply chain collaboration ($H_{3b}$). The direct effect of collaborative culture on supply chain collaboration is 0.24, significant at the level of 0.01.

**Table 7.18** Structural modeling results—indirect effects

| Hypotheses | Relationship | Direct | Indirect | Total | t-Value of indirect effect |
|---|---|---|---|---|---|
| $H_1$ | IR → IA | 0.26 | – | 0.26 | |
| $H_2$ | IA → SC | 0.37 | – | 0.37 | |
| $H_{3a}$ | CC → IA | 0.30 | – | 0.30 | |
| $H_{3b}$ | CC → SC | 0.24 | 0.11 | 0.35 | 2.26 |
| $H_{4a}$ | TR → IA | 0.44 | – | 0.44 | |
| $H_{4b}$ | TR → SC | 0.41 | 0.16 | 0.57 | 2.74 |
| $H_5$ | SC → CA | 0.63 | – | 0.63 | |
| $H_6$ | SC → FP | 0.36 | 0.31 | 0.68 | 5.53 |
| $H_7$ | CA → FP | 0.49 | – | 0.49 | |

Collaborative culture also has an indirect positive effect on supply chain collaboration (path coefficient = 0.11 and t-value = 2.26, significant at the 0.05 level), resulting in a total effect of 0.35. This indirect effect is mediated by IOS appropriation. Collaborative culture facilitates the extent of IOS use among the supply chain partners, which further intensifies the level of collaborations among partners.

It was postulated that trust has a significant positive relationship with supply chain collaboration ($H_{4b}$). From the results, $H_{4b}$ is supported with the significant, direct positive effect (path coefficient = 0.41, t-value = 4.95) at the 0.01 level. The indirect effect of trust on supply chain collaboration (path coefficient = 0.16, t-value = 2.74) is also significant at the 0.01 level. This indirect effect is through IOS appropriation, which further amplifies the level of collaboration among supply chain partners. It confirms that trust has significant positive effect on supply chain collaboration both directly and indirectly.

It was also hypothesized that supply chain collaboration has a significant positive relationship with firm performance ($H_7$). From the results, the direct effect of supply chain collaboration on firm performance (path coefficient = 0.36, t-value = 5.09) is significant at the 0.01 level. Supply chain collaboration also has significant indirect impact on firm performance through collaborative advantage (path coefficient = 0.31, t-value = 5.53) at the 0.01 level. Therefore, the collaborative advantage among supply chain partners is a huge amplifier that can help partners to achieve synergies and create superior firm performance.

Overall, the data indicate strong causal relationships among the constructs proposed in the framework.

## 7.3.2 Discussion of Hypotheses Testing Results

The results of the study confirm that IT resources have a significant positive, direct impact on IOS appropriation ($H_1$). The results show that IT infrastructure provides a common foundation (e.g., hardware, software, communication technologies, and databases) for the delivery of business applications and services, and thus flexibly supports IOS use for different purposes, e.g., integration of business processes, open communication, and joint knowledge creation among supply chain partners. The technical and managerial expertise of IT staff and managers makes the different uses of IOS easier and more effective. The findings are in line with the results of previous studies (Piccoli and Ives 2005; Byrd and Turner 2000; Bharadwaj 2000; Ross et al. 1996). However, this study further demonstrates that IT resources not only increase the intensity of IOS use but also support different uses of IOS among supply chain partners (e.g., use for integration, communication, and knowledge creation).

Looking at the antecedents to supply chain collaboration, the results first provide insight into the effectiveness of IOS appropriation as facilitators in enhancing supply chain collaboration ($H_2$). The finding indicates that different IOS appropriations support diverse activities of supply chain collaboration in terms of process integration, collaborative communication, and joint knowledge creation, even though the underlying technologies are similar. IOS use for integration allows information sharing and joint planning and execution of electronically coupling business processes between partners. IOS use for communication enables frequent and two-way message flows. IOS use for intelligence facilitates joint decision making and joint knowledge creation by using shared data repository, data warehouse, and data mining tools. The findings echo Malone et al's (1987) different impacts of IT (e.g., electronic communication effects and electronic integration effects) on interorganizational relationship and further identify a third use of IOS for intelligence. Although IOS use has been studied in the context of interorganizational relationship (Subramani 2004; Grover et al. 2002; Saeed et al. 2005; Barua et al. 2004), this research contributes to the literature by providing a more accurate definition of IOS appropriation and studying its impact in the supply chain context.

As expected, the results support the hypotheses that collaborative culture has significant, positive impact on IOS appropriation ($H_{3a}$) and on supply chain collaboration ($H_{3b}$). It demonstrates that firms with collaborative culture (collectivism, long-term orientation, power symmetry, uncertainty avoidance) are more likely to use IOS to integrate business processes, promote communications, and jointly explore new knowledge. It also indicates that collaborative culture helps to create common goals, information sharing, and open interaction and contribute significantly to effective supply chain collaboration. The finding is consistent with previous studies that effective supply chain collaboration cannot over-rely on the use of technologies and its cultural environment has to be taken into consideration (Barratt 2004). Companies with collaborative culture believe that their goals and those of their partners can go together and thus can work well and contribute their best to the chain (Wong 1999). Collaborative culture influences supply chain collaboration directly as well as indirectly through IOS appropriation. This indirect path shows that the commitment to creating a collaborative culture leads to better IOS use ($H_{3a}$), which in turn enhances supply chain collaboration ($H_2$). Although culture has been studied in the literature (Kumar et al. 1998; Mohr and Nevin 1990; Bates et al. 1995; Wuyts and Geyskens 2005; Sheu et al. 2006), its impacts on IOS use and supply chain collaboration have not been empirically tested before. This study has made important contributions on this prediction.

Notably, the study finds that trust has significant, positive impact on IOS appropriation ($H_{4a}$) and on supply chain collaboration ($H_{4b}$) with the highest path coefficients of 0.44 and 0.41 respectively among the facilitating factors. Trust is critical prerequisite for effective IOS use and supply chain collaboration. Trust influences supply chain collaboration both directly and indirectly through IOS use. The results indicate that trust is the most influential enabler in the model to increase the level of IOS appropriation and supply chain collaboration. Support for these hypotheses closely parallels findings in the trust and interorganizational management literature (Kumar et al. 1998; Duffy and Fearne 2004; Son et al. 2005; Sheu et al. 2006) where trust has been shown to make firms more willing to share internal information with their partners and make collaboration easier and smoother. The lack of trust can create serious problems for effective implementation of technologies and collaboration practices. As discussed in the literature, the big challenge for supply chain collaboration is trust and communication, not technology (Barratt 2004; Sheu et al. 2006). The effective implementation of supply chain collaboration practices needs the existence of trust, commitment, and shared goals between partners. Lack of trust and commitment can kill the collaboration in a very short time (Zineldin 1998).

Supply chain collaboration and collaborative advantage were found to exhibit a statistically significant positive relationship ($H_5$). From the results, the path coefficient from supply chain collaboration to collaborative advantage is the highest (0.63) among all, indicating a strong relationship between them. To the author's best knowledge, the study represents the first of its kind in the supply chain literature to define and operationalize collaborative advantage (i.e., joint competitive advantage) and to empirically test its relationship with supply chain collaboration. The results strongly suggest that better collaboration among supply chain partners expand the gain pie due to synergy through complementary resources and collaborative processes (Jap 1999; Tanriverdi 2006; Simatupang and Sridharan 2005).

The last finding was related to firm performance. The results empirically confirm that well executed supply chain collaboration directly improves firm performance ($H_6$) and collaborative advantage also increases firm performance directly ($H_7$). Previous research links collaboration directly to firm performance (Duffy and Feane 2004; Stank et al. 2001; Tan et al. 1998) without explicitly considering any intermediate variable such as collaborative advantage. This is an important finding since there exists doubt among researchers and practitioners in the economic justification of supply chain collaboration, particularly in whether collaborative advantage can bring financial benefits to the focal firm. The statistical significance of Hypotheses 6 and 7 suggests that supply chain collaboration and collaborative advantage indeed, have a bottom-line influence on the firm performance. The implementation of supply chain collaboration will improve a firm's financial performances in the long run. The results also show that supply chain collaboration has a significant, positive, indirect influence on firm performance through collaborative advantage ($H_5$ and $H_7$). The significance of Hypotheses 5 and 6, together with Hypothesis 7, the indirect influence discussed, jointly explains the critical role of supply chain collaboration in achieving collaborative advantage and improving firm performance.

# References

Anderson, J. C. (1987). An approach for confirmatory measurement and structural equation modeling of organizational properties. *Management Science, 33*, 525–541.

Anterasian, C., Graham, J. L., & Money, R. B. (1996). Are U.S. managers superstitious about market share? *Sloan Management Review, 37*(4), 67–77.

Bagozzi, R. P., & Philipps, L. W. (1982). Representing and testing organizational theories: A holistic construal. *Administrative Science Quarterly, 27*, 459–489.

Barratt, M. (2004). Understanding the meaning of collaboration in the supply chain. *Supply Chain Management: An Internal Journal, 9*(1), 30–42.

Barua, A., Konana, P., Whinston, A. B., & Yin, F. (2004). An empirical investigation of net-enabled business value. *MIS Quarterly, 28*(4), 585–620.

Bates, K. A., Amundson, S. D., Schroeder, R. G., & Morris, W. T. (1995). The crucial interrelationship between manufacturing strategy and organizational culture. *Management Science, 41*(10), 1565–1580.

Bentler, P. M. (1990). Comparative fit indexes in structural models. *Psychological Bulletin, 107*(2), 238–246.

Bentler, P. M., & Bonnet, D. (1980). Significance tests and goodness of fit in analysis of covariance structures. *Psychological Bulletin, 88*, 588–606.

Bharadwaj, A. S. (2000). A resource based perspective on information technology capability and firm performance: An empirical investigation. *MIS Quarterly, 24*(1), 169–196.

Byrd, T. A., & Turner, D. E. (2000). Measuring the flexibility of information technology infrastructure: Exploratory analysis of a construct. *Journal of Management Information Systems, 17*(1), 167–208.

Byrne, B. M. (1989). *A primer of LISREL: Basic applications and programming for confirmatory factor analysis analytic models*. New York: Springer-Verlog.

Chau, P. (1997). Reexamining a model for evaluating information center success using a structural equation modeling approach. *Decision Sciences, 28*(2), 309–334.

Chin, W.W. (1998). Issues and opinion on structural equation modeling. MIS Quarterly, 22(1), vii–xvi.

Crawford, S., McCabe, S., Couper, M., & Boyd, C. (2002). *Improving response rates and data collection efficiencies. Proceedings of International Conference on Improving Surveys*. Copenhagen, Denmark, August 25–28.

Diamantopoulos, A. (1999). Export performance measurement: Reflective versus formative indicators. *International Marketing Review, 16*(6), 444–457.

Doll, W. J., Raghunathan, T., Lim, S. J., & Gupta, Y. P. (1995). A confirmatory factor analysis of the user information satisfaction instrument. *Information Systems Research, 6*(2), 177–188.

Duffy, R., & Fearne, A. (2004). The impact of supply chain partnerships on supplier performance. *International Journal of Logistics Management, 15*(1), 57–71.

Grover, V., Teng, J., & Fiedler, K. (2002). Investigating the role of information technology in building buyer-supplier relationships. *Journal of Association for Information Systems, 3*, 217–245.

Hair, J. F., Anderson, R. E., Tatham, R. L., & Black, W. C. (1995). *Multivariate data analysis with readings*. New York: Prentice-Hall Inc.

Heck, R. H. (1998). Factor analysis: Exploratory and confirmatory approaches. In G. A. Marcoulides (Ed.), *Modern methods for business research* (pp. 177–215). Mahwah, NJ: Lawrence Erlbaum Associates.

Jap, S. D. (1999). Pie-expansion efforts: Collaboration processes in buyer-supplier relationships. *Journal of Marketing Research, 36*(4), 461–476.

Jarvis, C. B., MacKenzie, S. B., & Podsakoff, P. M. (2003). A critical review of construct indicators and measurement model misspecification in marketing and consumer research. *Journal of Consumer Research, 30*(2), 199–218.

Joreskog, K. G., & Sorbom, D. (1986). *LISREL VI: Analysis of linear structural relationships by maximum likelihood, instrumental variables, and least squares methods.* Moorsville, IN: Scientific Software Inc.

Joreskog, K. G., & Sorbom, D. (1989). *LISREL 7 users' reference guide.* Chicago, IL: Scientific Software Inc.

Kumar, K., Van Dissel, H. G., & Bielli, P. (1998). The merchant of Prato revisited: Toward a third rationality of information systems. *MIS Quarterly, 22*(2), 199–226.

Laudon, K., & Laudon, J. (2004). *Management information systems: Managing the digital firm* (8th ed.). New Jersey: Prentice-Hall.

Litwinenko, A., & Copper, C. L. (1994). The impact of trust status on corporate culture. *Journal of Management in Medicine, 8*(4), 8–17.

Malone, T. W., Yates, J., & Benjamin, R. I. (1987). Electronic markets and electronic hierarchies. *Communications of the ACM, 30*(6), 484–497.

Marsh, H. W., & Hocevar, D. (1985). Application of confirmatory factor analysis to the study of self-concept: First-and higher-order factor models and their invariance across groups. *Psychological Bulletin, 97*(3), 562–582.

Mohr, J., & Nevin, J. R. (1990). Communication strategies in marketing channels: A theoretical perspective. *Journal of Marketing, 54*(4), 36–51.

Papke-Shields, K. E., Malhotra, M. J., & Grover, V. (2002). Strategic manufacturing planning systems and their linkage to planning system success. *Decision Sciences, 33*(1), 1–30.

Patnayakuni, R., Rai, A., & Seth, N. (2006). Relational antecedents of information flow integration for supply chain coordination. *Journal of Management Information Systems, 23*(1), 13–49.

Piccoli, G., & Ives, B. (2005). IT-dependent strategic initiatives and sustained competitive advantage: A review and synthesis of the literature. *MIS Quarterly, 29*(4), 747–776.

Ross, J. W., Beath, C. M., & Goodhue, D. L. (1996). Develop long-term competitiveness through IT assets. *Sloan Management Review, 38*(1), 31–42.

Saeed, K. A., Malhotra, M. K., & Grover, V. (2005). Examining the impact of interorganizational systems on process efficiency and sourcing leverage in buyer–supplier dyads. *Decision Sciences, 36*(3), 365–396.

Segars, A. H., & Grover, V. (1993). Re-examining perceived ease of use and usefulness: A confirmatory factor analysis. *MIS Quarterly, 17*(4), 517–525.

Segars, A. H., & Grover, V. (1998). Strategic information systems planning success: An investigation of the construct and its measurement. MIS Quarterly, 21, 139–163

Sheu, C., Yen, H. R., & Chae, D. (2006). Determinants of supplier-retailer collaboration: Evidence from an international study. *International Journal of Operations & Production Management, 26*(1), 24–49.

Simatupang, T. M., & Sridharan, R. (2005). An Integrative framework for supply chain collaboration. *International Journal of Logistics Management, 16*(2), 257–274.

Son, J., Narasimhan, S., & Riggins, F. J. (2005). Effects of relational factors and channel climate on EDI Usage in the customer-supplier relationship. *Journal of Management information Systems, 22*(1), 321–353.

Stank, T. P., Keller, S. B., & Daugherty, P. J. (2001). Supply chain collaboration and logistical service performance. *Journal of Business Logistics, 22*(1), 29–48.

Subramani, M. (2004). How do suppliers benefit from information technology use in supply chain relationships? *MIS Quarterly, 28*(1), 45–73.

Tan, K. C., Kannan, V. R., & Handfield, R. B. (1998). Supply chain management: Supplier performance and firm performance. *International Journal of Purchasing and Materials Management, 34*(3), 2–9.

Tanriverdi, H. (2006). Performance effects of information technology synergies in multibusiness firms. *MIS Quarterly, 30*(1), 57–77.

Vishwanath, V., & Mark, J. (1997). Your brand's best strategy. *Harvard Business Review, 75*(3), 123–129.

Williams, L. J., Edwards, J. R., & Vandenberg, R. J. (2003). Recent advances in causal modeling methods for organizational and management research. *Journal of Management, 29*(6), 903–936.

Wong, A. (1999). Partnering through cooperative goals in supply chain relationships. *Total Quality Management, 10*(4/5), 786–792.

Wuyts, S., & Geyskens, I. (2005). The formation of buyer–supplier relationships: Detailed contract drafting and close partner selection. *Journal of Marketing, 69*(4), 103–117.

Zineldin, M. (1998). Towards an ecological collaborative relationship management. *European Journal of Marketing, 32*(11/12), 1138–1164.

# Chapter 8
# Research and Managerial Insights

**Abstract** The preceding chapters have developed a framework of supply chain collaboration, its antecedents, and consequences, and presented methods for measuring these constructs and empirically testing relationships among them. In this chapter, we provide additional insights on the moderation effect of firm size on the relationship between supply chain collaboration and its consequences—collaborative advantage and firm performance in particular. In examining the moderator of firm size that set boundary conditions for the effects of supply chain collaboration and collaborative advantage, we found interesting results that collaborative advantage completely mediates the relationship between supply chain collaboration and firm performance for small firms while it partially mediates the relationship for medium and large firms. Also, managerial guidelines for forming collaboration and managing ongoing relationships are provided. This chapter discusses the moderation effect of firm size on the relationships between supply chain collaboration and its consequences, and provides (1) discussion of research findings and major contributions (2) implications for practitioners (3) limitations of the research, and (4) recommendations for future research.

## 8.1 Moderation Effect of Firm Size

While the three hypotheses of $H_5$, $H_6$, and $H_7$ proposed in the framework are supported, it is not clear whether these hypotheses hold across small, medium and large firms. That means, would the relationships remain significant or invariant across the three groups? To test the moderating effect of firm size, a multi-group analysis of structural invariance across firm sizes was conducted in LISREL (Joreskog and Sorbom 1996; Schumacker and Marcoulides 1998). Firms with less than 250, between 250 and 500, and greater than 500 employees are respectively classified as small (n = 64), medium (n = 58), and large (n = 89). We tested first for the baseline model (i.e., equal pattern, Model 1 in Table 8.1), then for equal

M. Cao and Q. Zhang, *Supply Chain Collaboration*,
DOI: 10.1007/978-1-4471-4591-2_8, © Springer-Verlag London 2013

**Table 8.1** The moderation effect of firm size using multi-group analysis in LISREL

| Models | $\chi^2$ | df | $\chi^2/$df | CFI | NNFI | RMSEA | Nested models | $\Delta\chi^2$ | $\Delta$df | Significance level |
|---|---|---|---|---|---|---|---|---|---|---|
| 1. Equal pattern | 528.85 | 303 | 1.75 | 0.89 | 0.87 | 0.104 | | | | |
| 2. Equal factor loadings | 564.22 | 329 | 1.71 | 0.89 | 0.88 | 0.102 | 2-1 | 35.37 | 26 | 0.104 |
| 3. Equal factor Loadings, factor correlations | 564.22 | 331 | 1.70 | 0.89 | 0.88 | 0.101 | 3-2 | 0 | 2 | 1.000 |
| 4. Equal factor loadings, factor correlations, measurement errors | 586.08 | 345 | 1.70 | 0.89 | 0.89 | 0.100 | 4-3 | 21.86 | 14 | 0.082 |
| 4a. SCC→CA | 598.29 | 347 | 1.72 | 0.88 | 0.88 | 0.102 | 4a-4 | 12.21 | 2 | 0.002 |
| 4b. CA→FP | 616.77 | 347 | 1.78 | 0.88 | 0.88 | 0.106 | 4b-4 | 30.69 | 2 | 0.000 |
| 4c. SCC→FP | 646.05 | 347 | 1.86 | 0.87 | 0.86 | 0.111 | 4c-4 | 59.97 | 2 | 0.000 |

factor loadings (i.e., equal lamda, Model 2 in Table 8.1), for equal factor inter-correlations (i.e., equal phi, Model 3 in Table 8.1), for equal measurement errors (i.e., equal theta-delta and theta-epsilon, Model 4 in Table 8.1), and finally for equal structural coefficients (i.e., equal gamma and beta, Model 4a, 4b, and 4c in Table 8.1).

The baseline model (Model 1 in Table 8.1) has normed $\chi^2$, CFI, NNFI, and RMSEA of 1.75, 0.89, 0.87, and 0.104 respectively. Although RMSEA scores a little bit above 0.10, considering other model fit indices using a comprehensive assessment, the model fits sufficiently well with data and thus acceptable. Each factor loading (lamda) was forced to be equal across the three groups yielding Model 2 in Table 8.1. The $\chi^2$ difference between Model 2 and Model 1 is 35.37 with d.f. of 26. This insignificant result (p = 0.104) shows that factor loadings appear to be invariant across the three groups.

Similarly, The $\chi^2$ difference between Model 3 and Model 2 is 0 with two degrees of freedom. This insignificant result (p = 1.000) shows that factor loadings and factor inter-correlations appear to be invariant across the three groups. The $\chi^2$ difference between Model 4 and Model 3 is 21.86 with 14 degrees of freedom. This insignificant result (p = 0.082) shows that factor loadings, factor inter-correlations, and measurement errors appear to be invariant across the three groups.

The next set of analysis involves testing the invariance of the structural coefficients. The $\chi^2$ difference between Model 4a ($\chi^2_{M4a}-\chi^2_{M4} = 12.21$ with two degrees of freedom and a p = 0.002, which is statistically significant). The significant result support $H_5$ that the level of association between SCC and CA is different across small, medium, and large firms.

Similarly, the $\chi^2$ difference between Model 4b and Model 4 ($\chi^2_{M4b}-\chi^2_{M4} = 30.69$ with two degrees of freedom and a p = 0.000) is statistically

**Table 8.2** Path coefficients and t-values by firm size

| Firm size | $H_5$ (SCC→CA) | | $H_6$ (SCC→FP) | | $H_7$ (CA→FP) | | Mediation of CA on the relationship of SCC and FP |
|---|---|---|---|---|---|---|---|
| | Path coefficient | t-value | Path coefficient | t-value | Path coefficient | t-value | |
| Small (N = 64) | 0.59[a] | 4.50 | 0.18 | 1.38[b] | 0.61 | 4.15 | Complete mediation |
| Medium (N = 58) | 0.50 | 3.64 | 0.43 | 3.69 | 0.52 | 4.32 | Partial mediation |
| Large (N = 89) | 0.78 | 6.79 | 0.42 | 3.03 | 0.43 | 3.02 | Partial mediation |

[a] Value is a standardized structural coefficient; [b] indicates a non-significant value at the level of 0.05

significant, indicating that the level of association between CA and FP is different across the three groups and thus $H_7$ is supported. The $\chi 2$ difference between Model 4c and Model 4 ($\chi^2_{M4b} - \chi^2_{M4} = 59.97$ with 2 degrees of freedom and a p = 0.000) is statistically significant, supporting $H_6$ and indicating that the level of association between SCC and FP is different across the three groups.

To further analyze the moderation effect of firm size, the standardized structural path coefficients for SCC→CA, CA→FP, and SCC→FP across small, medium, and large firms are shown in Table 8.2. These path coefficients are also plotted in Fig. 8.1. For the relationship of SCC→CA, the path coefficients are all significant for the three groups. The path coefficients for small firms (path coefficient = 0.59, t = 4.50) and medium firms (path coefficient = 0.50, t = 3.64) are similar while the path coefficient for large firms (path coefficient = 0.78, t = 6.79) are much higher. The results support that large firms are more effective in jointly creating value with their partners than small and medium ones. This is also demonstrated by the successful collaboration stories of large firms such as Procter & Gamble, Hewlett-Packard, IBM, and Dell with their partners.

For the association of CA→FP, all three path coefficients are significant, but the path coefficient decreases as firm size increases (path coefficients are 0.61, 0.52, and 0.43 respectively for small, medium, and large firms). The association of CA with FP is stronger for smaller firms than for larger firms because smaller firms are more likely to be highly focused and thus their performance depends on the joint benefits of the supply chain collaboration for their limited set of products (Hendricks and Singhal 2005). Thus, smaller firms are more effective in improving their performance by appropriating the relational rents than larger ones. This doesn't mean that smaller firms get more returns in dollars since path coefficients are completely standardized. Smaller firms get more returns relative to their firm size than larger firms. This is consistent with the relative firm size hypothesis (i.e., value creation expressed as a percentage of firm size) that smaller firms get more abnormal returns from alliances than larger firms but their gains in dollar value might be similar (Koh and Venkatraman 1991; Anand and Khanna 2000).

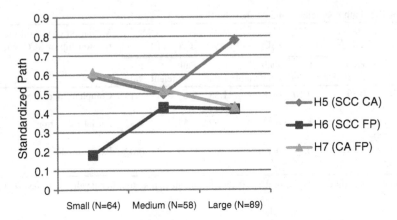

**Fig. 8.1** Plot of Standardized Path Coefficients of SCC→CA, CA→FP, SCC→FP Across Firm Size

For the relationship of SCC→FP, the path coefficient for small firms are insignificant (path coefficient = 0.18, t = 1.38) at the level of 0.05 while the path coefficients for both medium (path coefficient = 0.43, t = 3.69), and large firms (path coefficient = 0.42, t = 3.03) are significant and similar. This is in line with the extended resource based view that inter-firm learning and knowledge are easier to be transferred and utilized in other related business areas and thus generate more spillover rents for medium and large firms than small firms. This also echoes the relatedness hypothesis that larger firms can enjoy advantages of scale and scope economies, efficiency in resource allocation, and opportunities to use technical and managerial skills learned from partners in its related businesses (Koh and Venkatraman 1991; Ramaswami et al. 2009).

Based on the ERBV (Lavie 2006), collaborative advantage can be understood as a function of the combined value and rarity of all shared resources among supply chain partners (i.e., relational rents) while competitive advantage of a firm depends on the total value and rarity of the firm's own shared and non-shared resources (i.e., internal rents) and resources interactions with partners (i.e., appropriated relational rents and spillover rents). When collaboration is formed, each partner endows a subset of its resources to the collaboration with the expectation of generating common benefits from the shared resources of both firms (Lavie 2006).

According to the relational view and the ERBV (Dyer and Singh 1998; Lavie 2006), the link of SCC→CA (i.e., $H_5$) focuses on the joint value creation process. Firms generate common benefits (i.e., relational rents) through supply chain collaboration. The link of CA→FP (i.e., $H_7$) focuses on the value appropriation process. Firms improve their performance by appropriating relational rents. The direct link of SCC→FP (i.e., $H_6$) focuses on the spillover rents (and internal rents) that generate private benefits to the focal firm, which is not related to collaborative advantage. Such private benefits directly impact firms' performance.

In examining the moderator of firm size that set boundary conditions for SCC and CA effects, we found interesting results shown in Table 8.2. For small firms, the direct relationship between SCC and FP is insignificant (path coefficient = 0.18, t = 1.38). Both SCC→CA (path coefficient = 0.59, t = 4.50) and CA→FP (path coefficient = 0.61, t = 4.15) are significant. The results show that collaborative advantage completely mediate the relationship between SCC and FP for small firms. For medium firms, the path coefficients for SCC→CA, CA→FP, and SCC→FP are all significant so the collaborative advantage partially mediate the relationship between SCC and FP. Similarly, for large firms, all three path coefficients are significant so that CA also partially mediate the relationship between SCC and FP.

These results echo the literature acknowledging both common benefits and private benefits (Hamel 1991; Khanna et al. 1998; Lavie 2006). For small firms, SCC helps them jointly create value with their partners but their interfirm learning and joint knowledge are difficult to internalize or transfer to other business areas due to their small scale and scope of operations. Thus, SCC can cause little spillover effect that generates private benefits. In other words, SCC can achieve almost no benefits that are not related to the collaborative advantage, and thus it has no direct impact on firm performance. So the impact of SCC on firm performance is completely mediated by collaborative advantages for small firms.

For medium and large firms, it is easier for them to internalize and transfer what they have learned from supply chain collaboration to other related business areas due to their diversified businesses and large scope of operations. These spillover effects generate private benefits and thus SCC directly impact firm performance as well as indirectly improve firm performance through collaborative advantages by jointly creating value with their partners. This explains the partial mediation effect of collaborative advantage on the relationship between SCC and FP.

## 8.2 Research Contribution

In the past decade, there has been a need for firms to look outside their organizations for opportunities to collaborate with partners to ensure that the supply chain is both efficient and responsive to dynamic market needs. The role of information and associated technologies in facilitating and enabling supply chain collaboration has been stressed (Balakrishnan and Geunes 2004). However, knowledge of supply chain collaboration has been obscured by the vague terms of integration or partnership (Goffin et al. 2006; Frohlich and Westbrook 2001; Narasimhan and Kim 2002; Chen and Paulraj 2004; Petersen et al. 2005; Das et al. 2006; Deveraj et al. 2007; Zhao et al. 2008; Van der Vaart and van Donk 2008; Flynn et al. 2010) and fragmented studies focusing on a small number of factors (Sheu et al. 2006).

This study heeds the research calls by attempting to answer the following important research questions: (1) What are the key dimensions of supply chain collaboration? (2) What are the key dimensions of IOS appropriation? (3) What are the key dimensions of collaborative advantage? (4) What roles do the IT resources, IOS appropriation, culture, and trust play in improving supply chain collaboration? (5) How does supply chain collaboration help achieve collaborative advantage and firm performance? (6) Does collaborative advantage completely mediate the relationship between supply chain collaboration and firm performance? Does firm size moderate these relationships?

This study has defined and operationalized supply chain collaboration as a set of comprehensive components, and has investigated the antecedents (e.g., IT resources, IOS appropriation, culture, trust) and consequences (e.g. collaborative advantage and firm performance) of supply chain collaboration. Using a large-scale Web-based survey, 211 useful responses were collected from top management and executives. The proposed model was tested using structural equation modeling methodology.

To the author' best knowledge, the study represents one of the first large-scale empirical efforts to provide preliminary insights into the antecedents to, and the consequences of, supply chain collaboration. This study has made contributions to our understanding of IOS enabled supply chain collaboration, one of the most complex and challenging aspects of supply chain management, in a number of ways.

First, the research has provided a more accurate and comprehensive definition of supply chain collaboration. A significant amount of research has focused on the development of partnership models. What is lacking is a framework for accurately defining the extent of supply chain collaboration (Lambert et al. 1999). Previous definitions of supply chain collaboration put focus on process integration and largely ignore the components of communication and knowledge creation (Simatupang and Sridharan 2004; Nyaga et al. 2010; Verdecho et al. 2012; Fawcett et al. 2012). The current study has identified a comprehensive set of seven interconnecting dimensions that make up of effective supply chain collaboration: information sharing, goal congruence, decision synchronization, incentive alignment, resource sharing, collaborative communication, and joint knowledge creation.

These seven components in concert are sufficient and necessary to define the collaborative efforts. Collaborative efforts could include exchange of data about forecast, sales, stock levels, and delivery schedules, sharing of cost, establishing improvement teams together, mutual involvement in new product design, delivering training programs and providing technical assistance to partners, just-in-time delivery practices, development of logistics process, and definition of mutually shared performance metrics (Groves and Valsamakis 1998; Angeles and Nath 2001). Benefits of supply chain collaboration will be realized when all parties in the supply chain from suppliers to customers cooperate. Collaboration involves creating new value together rather than mere exchange (Kanter 1994).

Second, IOS appropriation has been defined in the study using dimensions appropriate for distinguishing between IOS use for integration, IOS use for communication, and IOS use for intelligence. This definition has emphasized the different purposes of IOS use. The three dimensions have their own focuses and play different roles in supply chain collaboration: integrating business processes, facilitating communication, and enhancing knowledge creation. The existent literature put excessive emphasis on IOS use for integration (Barua et al. 2004; Bensaou and Venkatraman 1995; Christiaanse and Venkatraman 2002; Manthou et al. 2004), however discount the other two dimensions. The important different roles identified in the definition of IOS appropriation have allowed researchers to accurately test the impact of IOS use on supply chain collaboration at the sub-construct level particularly.

Third, the research has emphasized the concept of collaborative advantage rather than competitive advantage. Collaborative advantage resides not within an individual firm, but across a firm's boundaries via partnering (Jap 2001; Dyer and Singh 1998; Kanter 1994). It is the strategic benefits gained by a group of collaborative firms. Although the concept of collaborative advantage is modestly discussed in the literature (Jap 2001; Kanter 1994), a reliable and valid operationalization of the concept has never been done to the author's best knowledge. This research has defined and operationalized collaborative advantage as five components: process efficiency, offering flexibility, business synergy, quality, and innovation. The operationalization of the concept facilitates further empirical research efforts. The collaborative advantage created by supply chain collaboration is undoubtedly an interesting research issue. Jointly creating the common pace of information sharing, replenishment, and supply synchronization in a supply chain can reduce excess inventory, avoid the costly bullwhip effect, enhance business synergy, improve quality, provide offering flexibility, and increase joint innovation.

Fourth, this research has provided a theoretical framework that identifies characteristics, antecedents, and consequences of IOS enabled supply chain collaboration. The conceptual model was built based on the review of a wide range of literature, incorporating appropriate features of interorganizational models from different perspectives (i.e., transaction cost economics, resource-based view, resource dependence theory, social exchange theory, trust-based rationalism, and knowledge perspective). By blending multiple theoretical perspectives, a full-round picture of supply chain collaboration has been painted. The framework has provided a foundation for future research. The framework can be used to study both collaboration formation and ongoing collaboration evaluation and maintenance to further enrich the collaboration theory.

Fifth, the study has developed valid and reliable instruments for supply chain collaboration and related constructs including: (1) IT resources, (2) IOS appropriation, (3) collaborative culture, (4) trust, (5) supply chain collaboration, and (6) collaborative advantage. These measures are useful to researchers who are interested in evaluating causes and effects of collaboration among supply chain partners. All the scales have been tested through rigorous statistical methodologies including pre-test, pilot-test using Q-sort method, confirmatory factor analysis, unidimensionality, reliability, and the validation of second-order construct. All the scales are shown to meet the requirements for reliability and validity and thus can be used in future research. The accurate definitions and measures of supply chain collaboration and related constructs has provided a rich and structured under-standing of what occurs in a supply chain or network. They also facilitate empirical research efforts because the relationships among constructs can be better captured with better definitions and measures. Good definitions and measures can provide analytical consistency that enables greater sharing and comparison of different research results.

Sixth, this study has provided strong evidences supporting the proposed hypotheses regarding IT resources, IOS appropriation, and supply chain collabo-ration. The study results suggest the critical roles of IOS in achieving better supply chain collaboration. Technologies can move collaboration to a closer to real-time basis for exchanging and utilizing shared information (Barratt 2004). Web-based digital applications enable supply chain collaboration to be carried out in a more fluid and interconnected inter-enterprise environment and products and services to be delivered at an Internet speed.

However, technology solutions are only part of the answers to improved supply chain collaboration. The findings of this study have also provided empirical sup-ports that demonstrate collaborative culture and trust have significant impact on supply chain collaboration directly and indirectly through IOS appropriation. IT resources and use are facilitating factors to achieving better supply chain collab-oration; however technologies alone are not sufficient. Managers have to make efforts to create a collaborative culture and a trust atmosphere to make supply chain collaboration effective. The study has contributed to the theory by incor-porating collaborative culture, trust in addition to IOS appropriation. This is an important contribution because it moves the theory beyond a transaction focus.

Seventh, the study has provided empirical evidence of the performance implications of collaboration, which have not been adequately addressed in the extant literature. In this fashion, the study has answered the calls of researchers who have stressed the need for empirical research that examines the outcomes of collaboration (Jap 1999; Wong 1999), the collaborative advantage and firm per-formance in particular. The research results have highlighted the critical role of supply chain collaboration and the amplifying role of collaborative advantage in achieving firms' performance. A higher level of supply chain collaboration directly leads to a higher level of firm performance. Supply chain collaboration helps achieve collaborative advantage, which in turn leads to a higher level of firm performance. The study has contributed to the growing literature on the role of collaboration in creating synergies and collaborative advantage.

Eighth, this study extends the theory of co-opetition, presented as a mixture of cooperation and competition, from individual firm context to the supply chain context. The theory of co-opetition asserts that players can benefit when they cooperate, and the sum of what is gained by all players is larger than the sum of what the players gain by acting alone (Brandenburger and Nalebuff 1996; Zineldin 1998; Jap 1999). Supply chain partners collaborate to achieve synergy effects and collaborative advantage and compete against other chains to gain competitive advantage. This is a partnership-based win–win situation. Based on resource dependence theory, firms try to get more resources and make themselves less dependent on their environment. In the supply chain context, the study findings suggest that firms in a supply chain should share resources and cooperate with their partners and make the whole chain less dependent on the environment. As such, the supply chain as a whole is more competitive.

Lastly, the study has also generated new insight for analyzing supply chain collaboration using different theories including resource-based view, trust-based rationalism, and knowledge perspective. It has brought the application of these theories in the individual firm context to the supply chain context. Resource-based view, trust-based rationalism, and knowledge perspective have been used to explain the phenomena of supply chain collaboration as supplements to TCE. TCE has been criticized for its sole focus on cost minimization and ignoring other important behavioral variables such as trust, power, and culture (Duffy and Fearne 2004). Resource-based view emphasizes the importance of resource complementarity for collaboration. While a firm's resources are not strategic when examined in isolation, as a system of complements, they become strategic when bundled with partners' complementary resources. Trust-based rationalism makes a behavioral assumption of trustworthiness and believes continuing collaboration is based on trust rather than on monitoring and control mechanism. Knowledge perspective regards the collaboration as a partner-enabled market knowledge creation process.

## 8.3 Managerial Implications

In addition to the theoretical contributions of the study, there are some practical implications that can be inferred.

First, as today's competition is no longer between firms but between supply chains, firms are facing critical challenges of how to collaborate well with their partners to improve performance. One of the key messages from this study is to reinforce the assertion that "to be an effective competitor in today's global market requires one to be an effective collaborator" (Morgan and Hunt 1994). An effective supply chain collaboration should have the following cornerstones: (1) Taking a relationship perspective on the collaboration and sharing a common vision with your partners; (2) Using technologies as means of co-creating values and building deeper relationships; (3) Creating a collaborative culture with a long-term orientation to work together and developing a philosophy that all partners in the chain are equal; (4) Creating a mutual trust environment for coordinating resources effectively and flexibly to achieve synergy advantages. Managers could plan and manage collaborations with their supply chain partners based on the above prescriptions. These cornerstones can be used for managers to determine whether the failure of collaboration is caused by improper execution of certain practices or by a poor assessment of the drivers and facilitator, thus further help to identify the most proper way to enhance supply chain collaboration.

Second, the definition and measures of supply chain collaboration as seven important elements can help managers to define specific actions to be taken collaboratively to improve shared supply chain processes that benefit all members. The definition and measurements can serve as a powerful tool for managers to form effective collaborative relationships. It can help firms to minimize the chance of collaboration failure by addressing these seven key dimensions before entering the collaborative relationship.

The measurements developed for supply chain collaboration are not only useful for managers to form a good relationship but also useful for managers to evaluate and maintain ongoing collaborative partnerships with supply chain partners. They can be used to measure and monitor the level of collaboration among supply chain partners and benchmark the performance of a supply chain. Supply chain partners need to develop joint mechanisms for evaluating the collaboration based on characteristics of a strong relationship and communication flows. There are many ways of assessing the strength of the relationship. For example, quarterly formal evaluation can be used to identify possible problems before they become major concerns. Communication flows can be improved through exchange programs or on-site visits.

Instruments for supply chain collaboration can also be used as a segmentation tool (Lambert et al. 1999). A firm can segment its supplier or customer base based on the proper degree of collaboration and tailor its purchasing and marketing strategies accordingly.

Third, firms are increasingly making investments on information technologies in the hope for improving supply chain collaboration and financial performance. There are actually conflicting evidences showing the potential benefits from IOS use. The findings of this study assure the practitioners that IOS can be deployed in different ways (e.g., for integration, communication, and intelligence) to enhance supply chain collaboration and further improve firm performance. As firms are shifting to do business on the electronic platforms, the Web-based technologies has become a foundation of doing business and their effective use has become a necessary condition for any firms to survive in collaborating with their supply chain partners.

Fourth, supply chain collaboration has proved difficult to implement because there has been an over-dependence on technology, an overlooked role of collaborative culture, and a lack of trust between partners (Barratt 2004). If managers of partner organizations are aware of these factors, they may be in a better position to manage collaboration between themselves and their partners. Managers need to be less obsessed with the technology and focus more on the social context (e.g. collaborative culture and trust) in which the technology exists. Managers need to establish a collaborative culture and a trust environment for effective collaboration.

The study results demonstrated that collaborative culture is prerequisite for the development of an environment in which supply chain collaboration can occur. When there is a lack of collaborative culture, supply chain collaboration is likely to decrease, and collaborative advantage will be adversely affected. Collaborative culture helps firms to overcome overwhelmingly pursuing individual firm's benefits. If necessary, a firm may need to change its culture—a movement away from an adversary relationship to one that is based on collectivism and long-term orientation.

Firms also need to learn to trust each other when they enter a collaborative relationship since trust is the cornerstone of collaborative long-term relationship (Sirdeshmukh et al. 2002). A firm needs to let its partners know that it is making an effort to develop and/or maintain high levels of trust and that there is little chance of opportunistic behavior. While opportunistic behavior may be individually rational for a partner, it is not collectively optimal (Hill 1990). Even though a partner may be able to get away with limited acts of opportunism, it is likely to indirectly influence the supply chain collaboration and collaborative advantage.

Fifth, the study found that effective supply chain collaboration leads to collaborative advantage and better firm performance. The relationship implies that, in order for a supply chain as a whole to perform well, firms should try to create a win–win situation that all participants collaborate to achieve business synergy and compete with other chains. Normally competitive expectations lead individual firms to promote their own interests at the expenses of others. This is very insidious for collaboration and it will worsen and destroy the relationships. Long-term relationships such as supply chain collaboration have to be motivated by the mutuality of intent, goal congruence, and benefit sharing (Wong 1999; Tuten and Urban 2001). Thus, managers need to align goals and benefits with supply chain partners for creating collaborative advantage. Such collaborative advantage indeed directly increases the financial performance for each partner in the chain. In addition, collaboration might also cause the increased cost of coordination and inflexibility (Das et al. 2006) so that managers need to strike the balance and find an optimal level of collaborative efforts for their firms.

Sixth, the model development and empirical testing presented in the study move our understanding of supply chain collaboration a step forward. They provide important guidance for managers to achieve better partnership formation, management, and outcomes. Collaboration is good, but firms must invest efforts to make it work. Collaboration fails largely because it is not well executed (Lambert et al. 1999).

True supply chain collaboration requires an understanding of each member's requirements and capabilities to set up a clear vision for value co-creation. If individual firms in a supply chain have their own plan for their activities, these plans are doomed to failure because they fail to take into account supply chain partners' plans that will impact the outcome of a particular plan. Often firms in a supply chain suffer from poor communication and do not have a common performance measures in place. As such, partners have conflicting behaviors and the supply chain pulls in conflicting directions. Managers in a supply chain have to change their mindsets from competition only to co-opetition (i.e., combination of cooperation and competition).

The study finding also has implications for firms looking for good supply chain partners. Good partnering candidates have values and norms that match those of the searching firms. They have similar strategic goals, organizational processes, and operational styles. Managers need to select firms with compatible cultural orientation or firms that are willing to cultivate a collaborative culture. Managers need to evaluate how close the fit is between their firm and the potential partners before engaging in the supply chain formation. Not all supply chain relationships are candidates for evolution into long-term close partnerships (Tuten and Urban 2001).

When a firm deals with a situation where there is a lack of partner fit, the firm may feel that it is worthwhile reconsidering its own strategic goals and organizational processes in light of the potential benefits to be gained from ensuring the success of the supply chain. Such an approach would mean that one partner could change in some way to bring it more in line with another partner in terms of the latter's strategic goals and/or operational styles.

Seventh, an analysis of moderator effect of firm size shows the importance of inter-firm learning and knowledge transfer. A firm's ability to internalize and transfer its inter-firm learning and knowledge from collaboration depends on its absorptive capacity, which measures a firm's ability to identify, evaluate, assimilate, and exploit external knowledge (Cohen and Levinthal 1990). A firm's absorptive capacity enhances its learning from partners and eventually contributes to firm performance. Thus, the better the focal firm's absorptive capacity, the higher the spillover rents and the relational rents appropriated by the focal firm will be. Thus, managers should manage to improve their absorptive capacity and thus improve their firm performance. Especially for small firms, they need to adequately expand the scope of their operations to transfer inter-firm learning from collaboration.

Firms seek to internalize the resources and skills of their collaborative partners to improve their performance. When the firm and its partners pursue such objectives, the parties might engage in learning races or one party exploits the collaboration for its private benefits, which lead to an unintended leakage of rents with no synergistic value creation (Hamel 1991; Khanna et al. 1998; Lavie 2006). So managers need to be co-opetitive and balance the collaboration and competition. To collaborate well, firms need to work closely with partners to jointly create values that generate common benefits; to be competitive, firms might also actively invest in improvement of absorptive capacity and thus earn spillover rents that generate private benefits.

## 8.4 Limitations and Future Research

While the research has made significant contributions to research and practice, there are some limitations that need to be considered when interpreting the study findings.

First, because the number of observations (211) is limited, the constructs were not revalidated in this research by splitting the observations as training and validation samples. This needs to be addressed in the future research. New data may be collected to revalidate the measures developed here.

Second, a key respondent, namely the top manager, in an organization was elicited to respond to a set of complex issues on supply chain collaboration, culture, trust, IOS use, collaborative advantage, and firm performance, since the top management is arguably the most knowledgeable individual about those issues. This may introduce common-method bias. The stability of the findings needs to be tested by generating data using multiple informants from within the organization, and using knowledgeable members of the organization.

Third, the response rate of 6 %, even though comparable to similar studies, is considered low. A main reason of the low response rate is the length of questionnaire. Because of the time constraint of top managers, they are unlikely to participate in lengthy surveys. This issue can be addressed in the future research by reducing the number of items in the questionnaire.

Fourth, data collection on both sides of the manufacturer-supplier dyad would alleviate concern about biased assessments. However, while collecting information about the same relationship from both sides of the dyad is advocated, it is very difficult to carry out in practice due to the operational difficulty and an adequate sample size (Duffy and Fearne 2004).

This study has provided a useful starting point from which to examine the roles of IOS appropriation, collaborative culture, and trust in supply chain collaboration and has identified several variables of notable research and managerial significance. As a result, there are a number of interesting areas in which future research could be undertaken to good effect.

Since the usefulness of a measurement scale comes from its generalizability, future research should revalidate measurement scales developed in this research by using similar reference populations. Future research should also conduct factorial invariance tests. Generalizability of measurement scales can further be supported by factorial invariance tests. Using the instruments developed in this research, one may test for factorial invariance across different organization size, across organizations with different supply chain structure (such as organization's position in the supply chain, channel structure, and so on), and across industries. For example, an analysis of supply chain collaboration and its related constructs by industry would be very beneficial. Examining how they are used across different industries and what are the most common level of supply chain collaboration in each industry would help identify any industry-specific bias toward or against supply chain collaboration.

Future research should apply multiple methods to obtain data. The use of a single respondent to represent what are supposed to supply chain wide variables may generate some inaccuracy and more than the usual amount of random error. Future research should try to use multiple respondents from each participating firm as an effort to enhance reliability of research findings. More insights will be gained by collecting information from both sides of the manufacturer-supplier dyad rather than just from one organization. Once a construct is measured with multiple methods, random error and method variance may be assessed using a multitrait-multimethod approach.

Future research should examine the hypothesized structural relationships across industries. Assuming an adequate sample size in each industry, structural analysis may be done by industry. This would reveal either industry-specific structural relationships or invariance of structural relationships across industries. The same hypothesized structural relationships across countries can also be tested in the future research. This will allow the comparison of the level of collaboration among supply chain partners across countries, the identification of country-specific facilitating and inhibiting factors, and the generalization of common collaboration and outcome factors across countries.

Future studies can also examine the proposed relationships by incorporating some contextual variables into the model, such as organizational size and production systems. For example, it will be interesting to investigate how supply chain collaboration differs across organization size. It will also be interesting to examine the impact of production systems (e.g., make-to-order and make-to-stock) on supply chain collaboration and performance.

In this study, composite measures are used to represent each construct, and only the construct-level structural model is tested using LISREL. However, the nature of relationships among sub-constructs across different variables will be more interesting. For example, what components of collaborative culture have more impact on supply chain collaboration? What differing roles of three components of IOS appropriation on supply chain collaboration? What dimensions of supply chain collaboration has more impact on collaborative advantage? By assessing these relationships at the sub-construct level, many alternative models can be explored and the findings will be more useful for decision makers.

While this study provided important insights into the determinants of supply chain collaboration, and of collaborative advantage, it did not shed much light on the change processes involved in the supply chain collaboration since time, and changes over time, were not explicitly modeled. However, research is needed at this level since supply chain partners learn from ongoing relationships and they modify business practices to better meet each other's needs to ensure the relationship remains adaptable and valuable (Min et al. 2005). In the future, other research designs such as longitudinal study and experimentation research can be conducted to help determine how collaboration-related factors and relationships change over time.

The model developed in the study does not purport to represent all the possible antecedents of supply chain collaboration. Future research can expand the current theoretical framework by incorporating new constructs. For example, one might include e-business and IS strategies into the existing framework.

# References

Anand, B., & Khanna, T. (2000). Do firms learn to create value? The case of alliances. *Strategic Management Journal, 21*, 295–315.

Angeles, R., & Nath, R. (2001). Partner congruence in electronic data interchange (EDI) enabled relationships. *Journal of Business Logistics, 22*(2), 109–127.

Balakrishnan, A., & Geunes, H. (2004). Collaboration and coordination in supply chain management and e-commerce. *Production and Operations Management, 13*(1), 1–2.

Barratt, M. (2004). Understanding the meaning of collaboration in the supply chain. *Supply Chain Management: An Internal Journal, 9*(1), 30–42.

Barua, A., Konana, P., Whinston, A. B., & Yin, F. (2004). An empirical investigation of net-enabled business value. *MIS Quarterly, 28*(4), 585–620.

Bensaou, M., & Venkatraman, N. (1995). Configurations of interorganizational relationships: A comparison between U.S. and Japanese automakers. *Management Science, 41*(9), 1471–1492.

Brandenburger, A., & Nalebuff, B. (1996). *Co-opetition.* New York: Currency Doubleday.

Chen, I. J., & Paulraj, A. (2004). Towards a theory of supply chain management: The constructs and measurements. *Journal of Operations Management, 22,* 119–150.

Christiaanse, E., & Venkatraman, N. (2002). Beyond SABRE: An empirical test of expertise exploitation in electronic channels. *MIS Quarterly, 26*(1), 15–38.

Cohen, W. D., & Levinthal, D. A. (1990). Absorptive capacity: A new perspective on learning and innovation. *Administrative Science Quarterly, 35*(1), 128–152.

Das, A., Narasimhan, R., & Talluri, S. (2006). Supplier integration: Finding an optimal configuration. *Journal of Operations Management, 24,* 563–582.

Deveraj, S., Krajewski, L., & Wei, J. (2007). Impact of eBusiness technologies on operational performance: The role of production information integration in the supply chain. *Journal of Operations Management, 25,* 1199–1216.

Duffy, R., & Fearne, A. (2004). The impact of supply chain partnerships on supplier performance. *International Journal of Logistics Management, 15*(1), 57–71.

Dyer, J. H., & Singh, H. (1998). The relational view: Cooperative strategy and sources of interorganizational competitive advantage. *Academy of Management Review, 23*(4), 660–679.

Fawcett, S. E., Fawcett, A., Watson, B., & Magnan, G. (2012). Peeking inside the black box: toward an understanding of supply chain collaboration dynamics. *Journal of Supply Chain Management, 48*(1), 44–72.

Flynn, B. B., Huo, B., & Zhao, X. (2010). The impact of supply chain integration on performance: A contingency and configuration approach. *Journal of Operations Management, 28*(1), 58–71.

Frohlich, M. T., & Westbrook, R. (2001). Arcs of integration: An international study of supply chain strategies. *Journal of Operations Management, 19*(2), 185–200.

Goffin, K., Lemke, F., & Szwejczewski, M. (2006). An exploratory study of close supplier-manufacturer relationships. *Journal of Operations Management, 24*(2), 189–209.

Groves, G., & Valsamakis, V. (1998). Supplier-customer relationships and company performance. *International Journal of Logistics Management, 9*(2), 51–64.

Hamel, G. (1991). Competition for competence and interpartner learning within international strategic alliances. *Strategic Management Journal, 12,* 83–103.

Hendricks, K., & Singhal, V. (2005). Association between supply chain glitches and operating performance. *Management Science, 51*(5), 695–711.

Hill, C. W. (1990). Cooperation, opportunism, and the invisible hand: Implications for transaction cost theory. *Academy of Management Review, 15,* 500–513.

Jap, S. D. (1999). Pie-expansion efforts: Collaboration processes in buyer-supplier relationships. *Journal of Marketing Research, 36*(4), 461–476.

Jap, S. D. (2001). Perspectives on joint competitive advantages in buyer-supplier relationships. *International Journal of Research in Marketing, 18*(1/2), 19–35.

Joreskog, K. G., & Sorbom, D. (1996). *LISREL 8 User's reference guide.* Chicago: Scientific Software Inc.

Kanter, R.M. (1994). Collaborative advantage: The art of alliances. *Harvard Business Review* (pp. 96–108), July–August.

Khanna, T., Gulati, R., & Nohria, N. (1998). The dynamics of learning alliances: Competition, cooperation, and relative scope. *Strategic Management Journal, 19,* 193–210.

Koh, J., & Venkatraman, N. (1991). Joint venture formations and stock market reactions: An assessment in the information technology sector. *Academy of Management Journal, 34*(4), 869–892.

Lambert, D. M., Emmelhainz, M. A., & Gardner, J. T. (1999). Building successful logistics partnerships. *Journal of Business Logistics, 20*(1), 165–181.

Lavie, D. (2006). The competitive advantage of interconnected firms: An extension of the resource-based view. *Academy of Management Review, 31*(3), 638–658.

Manthou, V., Vlachopoulou, M., & Folinas, D. (2004). Virtual e-Chain (VeC) model for supply chain collaboration. *International Journal of Production Economics, 87*(3), 241–250.

Min, S., Roath, A., Daugherty, P. J., Genchev, S. E., Chen, H., & Arndt, A. D. (2005). Supply chain collaboration: What's happening? *International Journal of Logistics Management, 16*(2), 237–256.

Morgan, R. M., & Hunt, S. D. (1994). The commitment-trust theory of relationship marketing. *Journal of Marketing, 58*(3), 20–38.

Narasimhan, R., & Kim, S. W. (2002). Effect of supply chain integration on the relationship between diversification and performance: Evidence from Japanese and Korean firms. *Journal of Operations Management, 20*(3), 303–323.

Nyaga, G., Whipple, J., & Lynch, D. (2010). Examining supply chain relationships: Do buyer and supplier perspectives on collaborative relationships differ? *Journal of Operations Management, 28*(2), 101–114.

Petersen, K., Handfield, R., & Ragatz, G. (2005). Supplier integration into new product development: Coordinating product, process, and supply chain design. *Journal of Operations Management, 23*(3/4), 371–388.

Ramaswami, S., Srivastava, R., & Bhargava, M. (2009). Market-based capabilities and financial performance of firms: Insights into marketing's contribution to firm value. *Journal of the Academy of Marketing Science, 37*, 97–116.

Schumacker, R., & Marcoulides, G. (1998). *Interaction and Nonlinear effects in structural equation modeling*. Mahwah: Lawrence Erlbaum.

Sheu, C., Yen, H. R., & Chae, D. (2006). Determinants of supplier-retailer collaboration: Evidence from an international study. *International Journal of Operations & Production Management, 26*(1), 24–49.

Simatupang, T. M., & Sridharan, R. (2004). A benchmarking scheme for supply chain collaboration. *Benchmarking: An International Journal, 11*(1), 9–30.

Sirdeshmukh, D., Singh, J., & Sabol, B. (2002). Consumer trust, value, and loyalty in relational exchanges. *Journal of Marketing, 66*(1), 15–37.

Tuten, T. L., & Urban, D. J. (2001). An expanded model of business-to-business partnership foundation and success. *Industrial Marketing Management, 30*(2), 149–164.

Van der Vaart, T., & Van Donk, D. (2008). A critical review of survey-based research in supply chain integration. *International Journal of Production Economics, 111*, 42–55.

Verdecho, M., Alfaro-Saiz, J., Rodriguez–Rodriguez, R., & Ortiz-Bas, A. (2012). A multi-criteria approach for managing inter-enterprise collaborative relationships. *Omega, 40*(3), 249–263.

Wong, A. (1999). Partnering through cooperative goals in supply chain relationships. *Total Quality Management, 10*(4/5), 786–792.

Zhao, X., Huo, B., Flynn, B. B., & Yeung, J. (2008). The impact of power and relationship commitment on the integration between manufacturers and customers in a supply chain. *Journal of Operations Management, 26*(3), 368–388.

Zineldin, M. (1998). Towards an ecological collaborative relationship management. *European Journal of Marketing, 32*(11/12), 1138–1164.

# Appendix
# Measurement Items, Criteria, and Questionnaire

## A: Measurement Items Entering Q-Sort

### IT Resources

*IT Infrastructure Flexibility*

Our systems are modular
Our systems are compatible
Our systems are scalable
Our systems are transparent
Our systems can handle multiple applications*
Our systems use commonly agreed IT standards

*IT Expertise*

Our IT staff has good technical knowledge
Our IT staff has the ability to quickly learn and apply new technologies as they become available
Our IT staff has the skills and experience to develop effective applications and systems
Our IT staff and managers have excellent business knowledge and deep understanding of business priorities and goals*
Our IT staff and managers understand our firm's technologies and business processes very well
Our IT staff and managers understand our firm's procedures and policies very well
Our IT staff and managers are knowledgeable about business strategy and business opportunities

M. Cao and Q. Zhang, *Supply Chain Collaboration*,
DOI: 10.1007/978-1-4471-4591-2, © Springer-Verlag London 2013

**IOS Appropriation**

*IOS Use for Integration*

Our firm and supply chain partners use IOS for integrating business functions across firms (e.g. design, manufacturing, and marketing)

Our firm and supply chain partners use IOS for joint forecasting, planning, and execution

Our firm and supply chain partners use IOS for order processing, invoicing and settling accounts

Our firm and supply chain partners use IOS for exchange of shipment and delivery information

Our firm and supply chain partners use IOS for managing warehouse stock and inventories

*IOS Use for Communication*

Our firm and supply chain partners use IOS for contacts about workflow coordination

Our firm and supply chain partners use IOS for conferencing

Our firm and supply chain partners use IOS for message services

Our firm and supply chain partners use IOS for frequent contacts

Our firm and supply chain partners use IOS for multiple channel communication

*IOS Use for Intelligence*

Our firm and supply chain partners use IOS for understanding trends in sales and customer preferences

Our firm and supply chain partners use IOS for storing, searching, and retrieving business information

Our firm and supply chain partners use IOS for deriving inferences from past events (e.g., process exceptions, patterns of demand shifts, what worked and what did not work)

Our firm and supply chain partners use IOS for combining information from different partners to uncover trends and patterns

Our firm and supply chain partners use IOS for interpreting information from different partners in multiple ways depending upon various requirements

**Collaborative Culture**

*Collectivism*

Our firm and supply chain partners are always jointly responsible for the successes and failures of our working relationships

Our firm considers it as the most normal thing that supply chain partners try to cooperate as much as possible

Close cooperation with supply chain partners is to be preferred over working independently

Our firm and supply chain partners focus on joint efforts with a feeling of "we are in this together"

## Long Term Orientation

Our firm wants and expects to have a long-term relationship with supply chain partners

Our firm believes that over the long run our relationship with supply chain partners is important to us

Our firm believes short-term inequities in the relationship would be balanced out by mutual benefits over the long term

Our firm is willing to make specific investments for long term relationships with supply chain partners

## Power Symmetry

Our firms believes that firms in the supply chain have an equal influence on each other

Our firms believes that firms in the supply chain that are in a powerful position should meet the needs of less powerful firms in mutually beneficial arrangements

Our firms believes that firms in the supply chain that are in a powerful position should have more to say in their relationships than their partners

Our firms believes that firms in the supply chain that are not in a powerful position should generally follow the will of their partners

## Uncertainty Avoidance

Uncertain situations in our supply chain are a threat to our firm

Our firm goes to great length to avoid uncertain situations in our supply chain

Our firm goes to great length to avoid unclear and ambiguous situations in our supply chain

Our firm tries to avoid risky situation in our supply chain

## Trust

### Credibility

Our supply chain partners are open and honest in dealing with us

Our supply chain partners are reliable

Our supply chain partners respect the confidentiality of the information they receive from us

Our supply chain partners usually keep the promises that they make to us

Our supply chain partners always provide accurate information

*Benevolence*

Our supply chain partners have made sacrifices for us in the past
Our supply chain partners are willing to provide assistance and support to us
    without exception
Our supply chain partners care for our welfare when making important decisions
When we share our problems with supply chain partners, we know that they will
    respond with understanding
We can count on supply chain partners to consider how their actions will affect us

## Supply Chain Collaboration

*Quality of Information Sharing*

Our firm and supply chain partners exchange relevant information
Our firm and supply chain partners exchange timely information
Our firm and supply chain partners exchange accurate information
Our firm and supply chain partners exchange complete information
Our firm and supply chain partners exchange confidential information
Our firm and supply chain partners exchange a variety of information*

*Goal Congruence*

Our firm and supply chain partners understand each other's needs and capabilities*
Our firm and supply chain partners have agreement on the goals of the supply
    chain
Our firm and supply chain partners have agreement on the importance of
    collaboration across the supply chain
Our firm and supply chain partners have agreement on the importance of
    improvements that benefit the supply chain as a whole
Our firm and supply chain partners agree that our own goals can be achieved
    through working towards the goals of the supply chain
Our firm and supply chain partners jointly layout collaboration implementation
    plans to achieve the goals of the supply chain

*Decision Synchronization*

Our firm and supply chain partners jointly plan on promotional events
Our firm and supply chain partners jointly develop demand forecasts
Our firm and supply chain partners jointly decide on optimal order quantity*
Our firm and supply chain partners jointly decide on inventory requirement
Our firm and supply chain partners jointly plan on product assortment
Our firm and supply chain partners jointly work out solutions

## Incentive Alignment

Our firm and supply chain partners co-develop systems to evaluate and publicize each other's performance (e.g. key performance index, scorecard, product/ service deliverables, and the resulting incentive)

Our firm and supply chain partners share costs and benefits

Our firm and supply chain partners share any risk that can occur in the supply chain

Our firm and supply chain partners share saving on reduced inventory costs

Our firm and supply chain partners have agreements on order changes*

The incentive for our firm is commensurate with our investment and risk

## Resource Sharing

Our firm and supply chain partners use cross-organizational teams frequently for process design and improvement

Our firm and supply chain partners dedicate personnel to manage the collaborative processes

Our firm and supply chain partners share technical support

Our firm and supply chain partners share equipments (e.g. computers, networks, machines)

Our firm and supply chain partners pool financial and non-financial resources (e.g. time, money, training, technology updates)

Our firm and supply chain partners make mutual resource investments dedicated to the relationships*

## Collaborative Communication

Our firm and supply chain partners have frequent meeting on a regular basis

Our firm and supply chain partners have open and two-way communication

Our firm and supply chain partners have informal communication

Our firm and supply chain partners have many different channels to communicate

Our firm and supply chain partners have high volume of coordination messages*

Our firm and supply chain partners influence each other's decisions through discussion rather than request

## Joint Knowledge Creation

Our firm and supply chain partners jointly search and acquire new and relevant knowledge

Our firm and supply chain partners jointly assimilate and apply relevant knowledge

Our firm and supply chain partners jointly understand customer needs

Our firm and supply chain partners jointly understand the market segments we serve*

Our firm and supply chain partners jointly understand new or emerging markets

Our firm and supply chain partners jointly understand intentions and capabilities of our competitors


## Collaborative Advantage

### Process Efficiency

Our firm with supply chain partners meets agreed upon costs per unit in comparison with industry norms

Our firm with supply chain partners meets productivity standards in comparison with industry norms

Our firm with supply chain partners meets on-time delivery requirements in comparison with industry norms

Our firm with supply chain partners meets inventory requirements (finished goods) in comparison with industry norms


### Offering Flexibility

Our firm with supply chain partners offers a variety of products and services efficiently in comparison with industry norms

Our firm with supply chain partners offers customized products and services with different features quickly in comparison with industry norms

Our firm with supply chain partners meets different customer volume requirements efficiently in comparison with industry norms

Our firm with supply chain partners has short customer response time in comparison with industry norms


### Business Synergy

Our firm and supply chain partners have integrated IT infrastructure and IT resources

Our firm and supply chain partners have integrated knowledge bases and know-how

Our firm and supply chain partners have integrated marketing efforts

Our firm and supply chain partners have integrated production systems

### Quality

Our firm with supply chain partners offers products that are highly reliable

Our firm with supply chain partners offers products that are highly durable

Our firm with supply chain partners offers high quality products to our customers

Our firm and supply chain partners have helped each other to improve product quality

*Innovation*

Our firm with supply chain partners introduces new products and services to market quickly

Our firm with supply chain partners has rapid new product development

Our firm with supply chain partners has time-to-market lower than industry average

Our firm with supply chain partners innovates frequently

Note: *Items were deleted after Q-sort.

## B: Cohen's Kappa and Moore and Benbasat Coefficients

The following example will to describe the Cohen's Kappa measure of agreement. Two judges independently classified a set of N components as either acceptable or rejectable. After the work was finished the following table was constructed:

| Judge 1 | | | | |
|---|---|---|---|---|
| Judge 2 | | Acceptable | Rejectable | Totals |
| | Acceptable | $X_{11}$ | $X_{12}$ | $X_{1+}$ |
| | Rejectable | $X_{21}$ | $X_{22}$ | $X_{2+}$ |
| | Totals | $X_{+1}$ | $X_{+2}$ | N |

$X_{ij}$ = the number of components in the ith row and jth column, for i,j = 1,2

The above table can also be constructed using percentages by dividing each numerical entry by N. For the population of components, the table will look like:

| Judge 1 | | | | |
|---|---|---|---|---|
| Judge 2 | | Acceptable | Rejectable | Totals |
| | Acceptable | $P_{11}$ | $P_{12}$ | $P_{1+}$ |
| | Rejectable | $P_{21}$ | $P_{22}$ | $P_{2+}$ |
| | Totals | $P_{+1}$ | $P_{+2}$ | 100 |

$P_{ij}$ = the percentage of components in the ith row and jth column

We will use this table of percentages to describe the Cohen's Kappa coefficient of agreement. The simplest measure of agreement is the proportion of components that were classified the same by both judges, i.e., $\Sigma_i P_{ii} = P_{11} + P_{22}$. However, Cohen suggested comparing the actual agreement, $\Sigma_i P_{ii}$, with the chance of agreement that would occur if the row and columns are independent, i.e., $\Sigma_i P_{i+}P_{+i}$. The difference between the actual and chance agreements, $\Sigma_i P_{ii} - \Sigma_i P_{i+}P_{+i}$, is the percent agreement above that which is due to chance. This difference can be standardized by dividing it by its maximum possible value, i.e., 100 % $- \Sigma_i P_i +$

$P_{+I} = 1 - \Sigma_i P_i + P_{+i}$. The ratio of these is denoted by the Greek letter kappa and is referred to as Cohen's Kappa.

Thus, Cohen's Kappa is a measure of agreement that can be interpreted as the proportion of joint judgment in which there is agreement after chance agreement is excluded. The three basic assumptions for this agreement coefficient are: (1) the units are independent, (2) the categories of the nominal scale are independents, mutually exclusive, and (3) the judges operate independently. For any problem in nominal scale agreement between two judges, there are only two relevant quantities:

$p_o$= the proportion of units in which the judges agreed

$p_c$= the proportion of units for which agreement is expected by chance

Like a correlation coefficient, $k = 1$ for complete agreement between the two judges. If the observed agreement is greater than or equal to chance $K <= 0$. The minimum value of $k$ occurs when $\Sigma P_{ii} = 0$, i.e.,

$$min(k) = \frac{-\Sigma_i(P_{i+} P_{+i})}{1 - \Sigma_i(P_{i+} P_{+i})}$$

When sampling from a population where only the total N is fixed, the maximum likelihood estimate of $k$ is achieved by substituting the sample proportions for those of the population. The formula for calculating the sample kappa ($k$) is:

$$k = \frac{N_i X_{ii} - \Sigma_i(X_{i+} X_{+i})}{N^2 - \Sigma_i(X_{i+} X_{+i})}$$

For kappa, no general agreement exists with respect to required scores. However, recent studies have considered scores greater than 0.65 to be acceptable (e.g. Vessey and Webber, 1984; Jarvenpaa 1989; Todd and Benbasat, 1991). Landis and Koch (1977) have provided a more detailed guideline to interpret kappa by associating different values of this index to the degree of agreement beyond chance. The following guideline is suggested:

| Value of kappa | Degree of agreement beyond chance |
| --- | --- |
| 0.76–1.00 | Excellent |
| 0.40–0.75 | Fair to good (moderate) |
| 0.39 or less | Poor |

A second overall measure of both the reliability of the classification scheme and the validity of the items was developed by Moore and Benbasat (1991). The method required analysis of how many items were placed by the panel of judges for each round within the target construct. In other words, because each item was included in the pool explicitly to measure a particular underlying construct, a measurement was taken of the overall frequency with which the judges placed items within the intended theoretical construct. The higher the percentage of items placed in the target construct, the higher the degree of inter-judge agreement across the panel that must have occurred.

Moreover, scales based on categories that have a high degree of correct placement of items within them can be considered to have a high degree of construct validity, with a high potential for good reliability scores. It must be emphasized that this procedure is more a qualitative analysis than a rigorous quantitative procedure. There are no established guidelines for determining good levels of placement, but the matrix can be used to highlight any potential problem areas. The following exemplifies how this measure works.

**Item Placement Scores**

| CONSTRUCTS | | ACTUAL | | | | | | |
|---|---|---|---|---|---|---|---|---|
| | | A | B | C | D | N/A | Total | % Hits |
| THEORETICAL | A | 26 | 2 | 1 | 0 | 1 | 30 | 87 |
| | B | 8 | 18 | 4 | 0 | 0 | 30 | 60 |
| | C | 0 | 0 | 30 | 0 | 0 | 30 | 100 |
| | D | 0 | 1 | 0 | 28 | 1 | 30 | 93 |

Item Placements: 120 Hits: 102 Overall "Hit Ratio": 85 %

The item placement ratio is an indicator of how many items were placed in the intended, or target, category by the judges. As an example of how this measure could be used, consider the simple case of four theoretical constructs with ten items developed for each construct. With a panel of three judges, a theoretical total of 30 placements could be made within each construct. Thereby, a theoretical versus actual matrix of item placements could be created as shown in the figure below (including an ACTUAL "N/A: Not Applicable" column where judges could place items which they felt fit none of the categories).

Examination of the diagonal of the matrix shows that with a theoretical maximum of 120 target placements (four constructs at 30 placements per construct), a total of 102 "hits" were achieved, for an overall "hit ratio" of 85 %. More important, an examination of each row shows how the items created to tap the particular constructs are actually being classified. For example, row C shows that all 30-item placements were within the target construct, but that in row B, only 60 % (18/30) were within the target. In the latter case, 8 of the placements were made in construct A, which might indicate the items underlying these placements are not differentiated enough from the items created for construct A. This finding would lead one to have confidence in scale based on row C, but be hesitant about accepting any scale based on row B. In an examination of off-diagonal entries indicate how complex any construct might be. Actual constructs based on columns with a high number of entries in the off diagonal might be considered too ambiguous, so any consistent pattern of item misclassification should be examined.

# C: Acronyms Used for Coding Items in Sub-Constructs

**IR**    **IT Resources**

     IRIFIT    Infrastructure Flexibility
     IRIEIT    Expertise

**IA**    **IOS Appropriation**

     IAIG    IOS Use for Integration
     IAIC    IOS Use for Communication
     IAIL    IOS Use for Intelligence

**CC**    **Collaborative Culture**

     CCCL    Collectivism
     CCLT    Long Term Orientation
     CCPS    Power Symmetry
     CCUA    Uncertainty Avoidance

**TR**    **Trust**

     TRCR    Credibility
     TRBN    Benevolence

**SC**    **Supply Chain Collaboration**

     SCIS    Quality of Information Sharing
     SCGC    Goal Congruence
     SCDS    Decision Synchronization
     SCIA    Incentive Alignment
     SCRS    Resource Sharing
     SCCM    Collaborative Communication
     SCKC    Joint Knowledge Creation

**CA**    **Collaborative Advantage**

     CAPE    Process Efficiency
     CAOF    Offering Flexibility
     CABS    Business Synergy
     CAPQ    Quality
     CAIN    Innovation

**FP**    **Firm Performance**

# D: Measurement Items After Q-Sort and Coding

**IT Resources**

*IT Infrastructure Flexibility*

IRIF1    Our systems are modular
IRIF2    Our systems are compatible

IRIF3    Our systems are scalable
IRIF4    Our systems are transparent
IRIF5    Our systems use commonly agreed IT standards

*IT Expertise*

IRIE1    Our IT staff has good knowledge of information technologies
IRIE2    Our IT staff has the ability to quickly learn and apply new information technologies as they become available
IRIE3    Our IT staff has the skills and experience to develop effective applications and systems
IRIE4    Our IT staff and managers understand our firm's technologies & business processes very well
IRIE5    Our IT staff and managers understand our firm's procedures and policies very well
IRIE6    *Our IT staff and managers are knowledgeable about our firm's business strategies, priorities, and opportunities*

## IOS Appropriation

*IOS Use for
Integration*

IAIG1    Our firm and supply chain partners use IOS for integrating business functions across firms (e.g. design, manufacturing, and marketing)
IAIG2    Our firm and supply chain partners use IOS for joint forecasting, planning, and execution
IAIG3    Our firm and supply chain partners use IOS for order processing, invoicing and settling accounts
IAIG4    Our firm and supply chain partners use IOS for exchange of shipment and delivery information
IAIG5    Our firm and supply chain partners use IOS for managing warehouse stock and inventories

*IOS Use for
Communication*

IAIC1    *Our firm and supply chain partners use IOS for workflow coordination*
IAIC2    Our firm and supply chain partners use IOS for conferencing
IAIC3    Our firm and supply chain partners use IOS for message services
IAIC4    Our firm and supply chain partners use IOS for frequent contacts
IAIC5    Our firm and supply chain partners use IOS for multiple channel communication

*IOS Use*
*for Intelligence*

IAIL1    Our firm and supply chain partners use IOS for understanding trends in
         sales and customer preferences
IAIL2    Our firm and supply chain partners use IOS for storing, searching, and
         retrieving business information
IAIL3    Our firm and supply chain partners use IOS for deriving inferences
         from past events (e.g., process exceptions, patterns of demand shifts,
         what worked and what did not work)
IAIL4    Our firm and supply chain partners use IOS for combining information
         from different sources to uncover trends and patterns
IAIL5    Our firm and supply chain partners use IOS for interpreting information
         from different sources in multiple ways depending upon various
         requirements

## Collaborative Culture

*Collectivism*

CCCL1    *Our firm and supply chain partners share responsibilities for the*
         *successes and failures of our working relationships*
CCCL2    Our firm considers it as the most normal thing that supply chain partners
         try to cooperate as much as possible
CCCL3    Close cooperation with supply chain partners is to be preferred by our
         firm over working independently
CCCL4    Our firm and supply chain partners focus on joint efforts with a feeling
         of "we are in this together"

*Long Term*
*Orientation*

CCLT1    Our firm wants and expects to have a long-term relationship with supply
         chain partners
CCLT2    Our firm believes that over the long run our relationships with supply
         chain partners are important to us
CCLT3    Our firm believes that short-term inequities in the relationship with
         supply chain partners would be balanced out by mutual benefits over
         the long term
CCLT4    Our firm is willing to make specific investments for long term
         relationships with supply chain partners

*Power Symmetry*

CCPS1    Our firm believes that firms in the supply chain have equal influence on each other
CCPS2    Our firm believes that firms in the supply chain that are in a powerful position should meet the needs of less powerful firms in mutually beneficial arrangements
CCPS3    Our firm believes that firms in the supply chain that are in a powerful position should have more to say in their relationships than their partners
CCPS4    Our firm believes that firms in the supply chain that are not in a powerful position should generally follow the will of their partners

*Uncertainty*
*Avoidance*

CCUA1    Uncertain situations in our supply chain are a threat to our firm
CCUA2    Our firm tries to avoid uncertain situations in our supply chain
CCUA3    Our firm tries to avoid unclear and ambiguous situations in our supply chain
CCUA4    Our firm tries to avoid risky situations in our supply chain

**Trust**

*Credibility*

TRCR1    Our supply chain partners are open and honest in dealing with us
TRCR2    Our supply chain partners are reliable
TRCR3    Our supply chain partners respect the confidentiality of the information they receive from us
TRCR4    Our supply chain partners usually keep the promises that they make to us
TRCR5    Our supply chain partners always provide accurate information

*Benevolence*

TRBN1    Our supply chain partners have made sacrifices for us in the past
TRBN2    Our supply chain partners are willing to provide assistance and support to us without exception
TRBN3    Our supply chain partners care for our welfare when making important decisions

TRBN4    When we share our problems with supply chain partners, we know that
         they will respond with understanding
TRBN5    We can count on supply chain partners to consider how their actions
         will affect us


**Supply Chain Collaboration**

*Quality of
Information
Sharing*

SCIS1    Our firm and supply chain partners exchange relevant information
SCIS2    Our firm and supply chain partners exchange timely information
SCIS3    Our firm and supply chain partners exchange accurate information
SCIS4    Our firm and supply chain partners exchange complete information
SCIS5    Our firm and supply chain partners exchange confidential information


*Goal
Congruence*

SCGC1    Our firm and supply chain partners have agreement on the goals
         of the supply chain
SCGC2    Our firm and supply chain partners have agreement on the
         importance of collaboration across the supply chain
SCGC3    Our firm and supply chain partners have agreement on the
         importance of improvements that benefit the supply chain as
         a whole
SCGC4    Our firm and supply chain partners agree that our own goals can
         be achieved through working towards the goals of the supply
         chain
SCGC5    Our firm and supply chain partners jointly layout collaboration
         implementation plans to achieve the goals of the supply chain


*Decision
Synchronization*

SCDS1    Our firm and supply chain partners jointly plan on promotional events
SCDS2    Our firm and supply chain partners jointly develop demand forecasts
SCDS3    *Our firm and supply chain partners jointly manage inventory*

SCDS4    Our firm and supply chain partners jointly plan on product assortment
SCDS5    Our firm and supply chain partners jointly work out solutions

*Incentive*
*Alignment*

SCIA1    Our firm and supply chain partners co-develop systems to evaluate and
            publicize each other's performance (e.g. key performance index,
            scorecard, and the resulting incentive)
SCIA2    *Our firm and supply chain partners share costs (e.g. loss on order*
            *changes)*
SCIA3    *Our firm and supply chain partners share benefits (e.g. saving on*
            *reduced inventory costs)*
SCIA4    Our firm and supply chain partners share any risks that can occur in the
            supply chain
SCIA5    The incentive for our firm is commensurate with our investment and
            risk

*Resource*
*Sharing*

SCRS1    Our firm and supply chain partners use cross-organizational teams
            frequently for process design and improvement
SCRS2    Our firm and supply chain partners dedicate personnel to manage the
            collaborative processes
SCRS3    Our firm and supply chain partners share technical supports
SCRS4    Our firm and supply chain partners share equipments (e.g. computers,
            networks, machines)
SCRS5    *Our firm and supply chain partners pool financial and non-financial*
            *resources (e.g. time, money, training)*

*Collaborative*
*Communication*

SCCM1    *Our firm and supply chain partners have frequent contacts on a regular*
            *basis*
SCCM2    Our firm and supply chain partners have open and two-way
            communication
SCCM3    Our firm and supply chain partners have informal communication

SCCM4   Our firm and supply chain partners have many different channels to communicate

SCCM5   Our firm and supply chain partners influence each other's decisions through discussion rather than request

*Joint*
*Knowledge*
*Creation*

SCKC1   Our firm and supply chain partners jointly search and acquire new and relevant knowledge

SCKC2   Our firm and supply chain partners jointly assimilate and apply relevant knowledge

SCKC3   *Our firm and supply chain partners jointly identify customer needs*

SCKC4   *Our firm and supply chain partners jointly discover new or emerging markets*

SCKC5   *Our firm and supply chain partners jointly learn the intentions and capabilities of our competitors*

## Collaborative Advantage

*Process*
*Efficiency*

CAPE1   *Our firm with supply chain partners meets agreed upon unit costs in comparison with industry norms*

CAPE2   Our firm with supply chain partners meets productivity standards in comparison with industry norms

CAPE3   Our firm with supply chain partners meets on-time delivery requirements in comparison with industry norms

CAPE4   Our firm with supply chain partners meets inventory requirements (finished goods) in comparison with industry norms

*Offering*
*Flexibility*

CAOF1   Our firm with supply chain partners offers a variety of products and services efficiently in comparison with industry norms

CAOF2   Our firm with supply chain partners offers customized products and services with different features quickly in comparison with industry norms

CAOF3   Our firm with supply chain partners meets different customer volume requirements efficiently in comparison with industry norms

CAOF4   *Our firm with supply chain partners has good customer responsiveness in comparison with industry norms*

*Business*
*Synergy*

CABS1    Our firm and supply chain partners have integrated IT infrastructure and IT resources
CABS2    Our firm and supply chain partners have integrated knowledge bases and know-how
CABS3    Our firm and supply chain partners have integrated marketing efforts
CABS4    Our firm and supply chain partners have integrated production systems

*Quality*

CAQL1    Our firm with supply chain partners offers products that are highly reliable
CAQL2    Our firm with supply chain partners offers products that are highly durable
CAQL3    Our firm with supply chain partners offers high quality products to our customers
CAQL4    Our firm and supply chain partners have helped each other to improve product quality

*Innovation*

CAIN1    Our firm with supply chain partners introduces new products and services to market quickly
CAIN2    Our firm with supply chain partners has rapid new product development
CAIN3    Our firm with supply chain partners has time-to-market lower than industry average
CAIN4    Our firm with supply chain partners innovates frequently

*Firm*
*Performance*

FP1      Market share
FP2      Growth of market share
FP3      Growth of sales
FP4      Return on investment
FP5      Growth in return on investment
FP6      Profit margin on sales
FP7      Overall competitive position

Note: Italicized items were reworded in Q-sort.

# E: Large-Scale Mail Survey Questionnaire

Supply chain partners refer to your primary partners that play a critical role in your firm and whose business fortune depends all or in part on the success of your firm. These include suppliers, contract manufacturers, subassembly plants, distribution centers, carriers, freight forwarder services, wholesalers, retailers, customers, and so on. Interorganizational systems (IOS) refer to the information technology applications that span firm boundaries.

Unless otherwise specifically requested, please use the following scale to answer each item:

| 1 | 2 | 3 | 4 | 5 | NA |
|---|---|---|---|---|---|
| Strongly Disagree | Disagree | Neutral | Agree | Strongly Agree | Not Applicable |

## Section 1. About IT Resources of Your Firm

The following statements describe the IT assets and capabilities in your firm. Please select the appropriate number to indicate the extent to which you agree or disagree with each statement as applicable to your firm.

### IT INFRASTRUCTURE FLEXIBILITY

Our systems (i.e., software, hardware, communication technologies, and database) ...

| | | | | | | |
|---|---|---|---|---|---|---|
| are modular | 1 | 2 | 3 | 4 | 5 | NA |
| are compatible | 1 | 2 | 3 | 4 | 5 | NA |
| are scalable | 1 | 2 | 3 | 4 | 5 | NA |
| transparent | 1 | 2 | 3 | 4 | 5 | NA |
| use commonly agreed IT standards | 1 | 2 | 3 | 4 | 5 | NA |

### IT EXPERTISE

| | | | | | | |
|---|---|---|---|---|---|---|
| Our IT staff has good knowledge of current information technologies | 1 | 2 | 3 | 4 | 5 | NA |
| Our IT staff has the ability to quickly learn and apply new information technologies as they become available | 1 | 2 | 3 | 4 | 5 | NA |
| Our IT staff has the skills and experience to develop effective applications and systems | 1 | 2 | 3 | 4 | 5 | NA |
| Our IT staff and managers understand our firm's technologies and business processes very well | 1 | 2 | 3 | 4 | 5 | NA |
| Our IT staff and managers understand our firm's procedures and policies very well | 1 | 2 | 3 | 4 | 5 | NA |
| Our IT staff and managers are knowledgeable about our business strategies, priorities, and opportunities | 1 | 2 | 3 | 4 | 5 | NA |

## Section 2. About the IOS Use in Your Supply Chain

The following statements describe IOS use in your supply chain. Please select the appropriate number to indicate the extent to which you agree or disagree with each statement as applicable to your firm.

### IOS USE FOR INTEGRATION

The extent of IOS use (e.g. EDI, ERP, MRP, CPFR, CRM, VMI, RFID) among supply chain partners for ...

| | | | | | | |
|---|---|---|---|---|---|---|
| integrating business functions across firms (e.g. design, manufacturing, and marketing) | 1 | 2 | 3 | 4 | 5 | NA |
| joint forecasting, planning, and execution | 1 | 2 | 3 | 4 | 5 | NA |
| order processing, invoicing and settling accounts | 1 | 2 | 3 | 4 | 5 | NA |
| exchange of shipment and delivery information | 1 | 2 | 3 | 4 | 5 | NA |
| managing warehouse stock and inventories | 1 | 2 | 3 | 4 | 5 | NA |

### IOS USE FOR COMMUNICATION

The extent of IOS use (e.g. email, video conferencing, electronic bulletin board, Intranet) among supply chain partners for ...

| | | | | | | |
|---|---|---|---|---|---|---|
| workflow coordination | 1 | 2 | 3 | 4 | 5 | NA |
| conferencing | 1 | 2 | 3 | 4 | 5 | NA |
| message services | 1 | 2 | 3 | 4 | 5 | NA |
| frequent contacts | 1 | 2 | 3 | 4 | 5 | NA |
| multiple channel communication | 1 | 2 | 3 | 4 | 5 | NA |

### IOS USE FOR INTELLIGENCE

The extent of IOS use (e.g. data mining/warehousing, OLAP, DSS, expert systems) among supply chain partners for ...

| | | | | | | |
|---|---|---|---|---|---|---|
| understanding trends in sales and customer preferences | 1 | 2 | 3 | 4 | 5 | NA |
| storing, searching, and retrieving business information | 1 | 2 | 3 | 4 | 5 | NA |
| deriving inferences from past events (e.g., process exceptions, patterns of demand shifts, what worked and what did not work) | 1 | 2 | 3 | 4 | 5 | NA |
| combining information from different sources to uncover trends and patterns | 1 | 2 | 3 | 4 | 5 | NA |
| interpreting information from different sources in multiple ways depending upon various requirements | 1 | 2 | 3 | 4 | 5 | NA |

## Section 3. About the Collaborative Culture in Your Firm

The following statements describe the beliefs and underlying values shared in your firm regarding appropriate business practices in the supply chain. Please select the number to indicate the extent to which you agree or disagree with each statement as applicable to your firm.

### COLLECTIVISM

| | | | | | | |
|---|---|---|---|---|---|---|
| Our firm and supply chain partners share responsibilities for the successes and failures of our working relationships | 1 | 2 | 3 | 4 | 5 | NA |
| Our firm considers it as the most normal thing that supply chain partners try to cooperate as much as possible | 1 | 2 | 3 | 4 | 5 | NA |
| Close cooperation with supply chain partners is to be preferred by our firm over working independently | 1 | 2 | 3 | 4 | 5 | NA |
| Our firm and supply chain partners focus on joint efforts with a feeling of "we are in this together" | 1 | 2 | 3 | 4 | 5 | NA |

### LONG TERM ORIENTATION

| | | | | | | |
|---|---|---|---|---|---|---|
| Our firm wants and expects to have a long-term relationship with supply chain partners | 1 | 2 | 3 | 4 | 5 | NA |
| Our firm believes that over the long run our relationships with supply chain partners are important to us | 1 | 2 | 3 | 4 | 5 | NA |
| Our firm believes that short-term inequities in the relationship with supply chain partners would be balanced out by mutual benefits over the long term | 1 | 2 | 3 | 4 | 5 | NA |
| Our firm is willing to make specific investments for long term relationships with supply chain partners | 1 | 2 | 3 | 4 | 5 | NA |

### POWER SYMMETRY

Our firm believes that firms in the supply chain ...

| | | | | | | |
|---|---|---|---|---|---|---|
| have equal influence on each other | 1 | 2 | 3 | 4 | 5 | NA |
| that are in a powerful position should meet the needs of less powerful firms in mutually beneficial arrangements | 1 | 2 | 3 | 4 | 5 | NA |
| that are in a powerful position should have more to say in their relationships than their partners | 1 | 2 | 3 | 4 | 5 | NA |
| that are not in a powerful position should generally follow the will of their partners | 1 | 2 | 3 | 4 | 5 | NA |

### UNCERTAINTY AVOIDANCE

| | | | | | | |
|---|---|---|---|---|---|---|
| Uncertain situations in our supply chain are a threat to our firm | 1 | 2 | 3 | 4 | 5 | NA |

| Our firm tries to avoid uncertain situations in our supply chain | 1 | 2 | 3 | 4 | 5 | NA |
|---|---|---|---|---|---|---|

| Our firm tries to avoid unclear and ambiguous situations in our supply chain | 1 | 2 | 3 | 4 | 5 | NA |
|---|---|---|---|---|---|---|

| Our firm tries to avoid risky situations in our supply chain | 1 | 2 | 3 | 4 | 5 | NA |
|---|---|---|---|---|---|---|

### Section 4. About Trust in Your Supply Chain

The following statements describe your firm's beliefs that supply chain partners will perform work and transactions according to your confident expectations, regardless of your ability to monitor them. Please select the number to indicate the extent to which you agree or disagree with each statement as applicable to your firm.

#### CREDIBILITY

Our supply chain partners ...

| are open and honest in dealing with us | 1 | 2 | 3 | 4 | 5 | NA |
|---|---|---|---|---|---|---|
| are reliable | 1 | 2 | 3 | 4 | 5 | NA |
| respect the confidentiality of the information they receive from us | 1 | 2 | 3 | 4 | 5 | NA |
| usually keep the promises that they make to us | 1 | 2 | 3 | 4 | 5 | NA |
| are capable of providing needed information | 1 | 2 | 3 | 4 | 5 | NA |

#### BENEVOLENCE

| Our supply chain partners have made sacrifices for us in the past | 1 | 2 | 3 | 4 | 5 | NA |
|---|---|---|---|---|---|---|
| Our supply chain partners are willing to provide assistance and support to us without exception | 1 | 2 | 3 | 4 | 5 | NA |
| Our supply chain partners care for our welfare when making important decisions | 1 | 2 | 3 | 4 | 5 | NA |
| When we share our problems with supply chain partners, we know that they will respond with understanding | 1 | 2 | 3 | 4 | 5 | NA |
| We can count on supply chain partners to consider how their actions will affect us | 1 | 2 | 3 | 4 | 5 | NA |

### Section 5. About Supply Chain Collaboration

The following statements describe your firm's long term partnership and collaboration with your supply chain partners. Please select the number to indicate the extent to which you agree or disagree with each statement as applicable to your firm.

#### QUALITY OF INFORMATION SHARING

Our firm and supply chain partners exchange ...

| relevant information | 1 | 2 | 3 | 4 | 5 | NA |
|---|---|---|---|---|---|---|
| timely information | 1 | 2 | 3 | 4 | 5 | NA |
| accurate information | 1 | 2 | 3 | 4 | 5 | NA |
| complete information | 1 | 2 | 3 | 4 | 5 | NA |
| confidential information | 1 | 2 | 3 | 4 | 5 | NA |

#### GOAL CONGRUENCE

Our firm and supply chain partners ...

| have agreement on the goals of the supply chain | 1 | 2 | 3 | 4 | 5 | NA |
|---|---|---|---|---|---|---|
| have agreement on the importance of collaboration across the supply chain | 1 | 2 | 3 | 4 | 5 | NA |
| have agreement on the importance of improvements that benefit the supply chain as a whole | 1 | 2 | 3 | 4 | 5 | NA |
| agree that our own goals can be achieved through working towards the goals of the supply chain | 1 | 2 | 3 | 4 | 5 | NA |
| jointly layout collaboration implementation plans to achieve the goals of the supply chain | 1 | 2 | 3 | 4 | 5 | NA |

#### DECISION SYNCHRONIZATION

Our firm and supply chain partners jointly ...

| plan on promotional events | 1 | 2 | 3 | 4 | 5 | NA |
|---|---|---|---|---|---|---|
| develop demand forecasts | 1 | 2 | 3 | 4 | 5 | NA |
| manage inventory | 1 | 2 | 3 | 4 | 5 | NA |
| plan on product assortment | 1 | 2 | 3 | 4 | 5 | NA |
| work out solutions | 1 | 2 | 3 | 4 | 5 | NA |

#### INCENTIVE ALIGNMENT

Our firm and supply chain partners ...

| share costs (e.g. loss on order changes) | 1 | 2 | 3 | 4 | 5 | NA |
|---|---|---|---|---|---|---|
| share benefits (e.g. saving on reduced inventory costs) | 1 | 2 | 3 | 4 | 5 | NA |
| co-develop systems to evaluate and publicize each other's performance (e.g. key performance index, scorecard, and the resulting incentive) | 1 | 2 | 3 | 4 | 5 | NA |
| share any risks that can occur in the supply chain | 1 | 2 | 3 | 4 | 5 | NA |
| The incentive for our firm is commensurate with our investment and risk | 1 | 2 | 3 | 4 | 5 | NA |

#### RESOURCE SHARING

Our firm and supply chain partners ...

| use cross-organizational teams frequently for process design and improvement | 1 | 2 | 3 | 4 | 5 | NA |
|---|---|---|---|---|---|---|
| dedicate personnel to manage collaborative processes | 1 | 2 | 3 | 4 | 5 | NA |
| share technical supports | 1 | 2 | 3 | 4 | 5 | NA |
| share equipments (e.g. computers, networks, machines) | 1 | 2 | 3 | 4 | 5 | NA |
| pool financial and non-financial resources (e.g. time, money, training) | 1 | 2 | 3 | 4 | 5 | NA |

#### COLLABORATIVE COMMUNICATION

Our firm and supply chain partners ...

| have frequent contacts on a regular basis | 1 | 2 | 3 | 4 | 5 | NA |
|---|---|---|---|---|---|---|
| have open and two-way communication | 1 | 2 | 3 | 4 | 5 | NA |
| have informal communication | 1 | 2 | 3 | 4 | 5 | NA |
| have many different channels to communicate | 1 | 2 | 3 | 4 | 5 | NA |
| influence each other's decisions through discussion rather than request | 1 | 2 | 3 | 4 | 5 | NA |

#### JOINT KNOWLEDGE CREATION

Our firm and supply chain partners jointly ...

| search and acquire new and relevant knowledge | 1 | 2 | 3 | 4 | 5 | NA |
|---|---|---|---|---|---|---|
| assimilate and apply relevant knowledge | 1 | 2 | 3 | 4 | 5 | NA |
| identify customer needs | 1 | 2 | 3 | 4 | 5 | NA |
| discover new or emerging markets | 1 | 2 | 3 | 4 | 5 | NA |
| learn the intentions and capabilities of our competitors | 1 | 2 | 3 | 4 | 5 | NA |

### Section 6. About Collaborative Advantage

The following statements describe the strategic benefits gained over competitors in the marketplace through supply chain partnering. Please select the appropriate number to indicate the extent to which you agree or disagree with each statement as applicable to your firm.

#### PROCESS EFFICIENCY

In comparison with industry norms, our firm with supply chain partners ...

| meets agreed upon unit costs | 1 | 2 | 3 | 4 | 5 | NA |
|---|---|---|---|---|---|---|
| meets productivity standards | 1 | 2 | 3 | 4 | 5 | NA |
| meets on-time delivery requirements | 1 | 2 | 3 | 4 | 5 | NA |
| meets inventory requirements (finished goods) | 1 | 2 | 3 | 4 | 5 | NA |

#### OFFERING FLEXIBILITY

In comparison with industry norms, our firm with supply chain partners ...

| | | | | | | |
|---|---|---|---|---|---|---|
| offers a variety of products and services efficiently | 1 | 2 | 3 | 4 | 5 | NA |
| offers customized products and services with different features quickly | 1 | 2 | 3 | 4 | 5 | NA |
| meets different customer volume requirements efficiently | 1 | 2 | 3 | 4 | 5 | NA |
| has good customer responsiveness | 1 | 2 | 3 | 4 | 5 | NA |

### BUSINESS SYNERGY
Our firm and supply chain partners have integrated ...

| | | | | | | |
|---|---|---|---|---|---|---|
| IT infrastructure and IT resources | 1 | 2 | 3 | 4 | 5 | NA |
| knowledge bases and know-how | 1 | 2 | 3 | 4 | 5 | NA |
| marketing efforts | 1 | 2 | 3 | 4 | 5 | NA |
| production systems | 1 | 2 | 3 | 4 | 5 | NA |

### QUALITY

| | | | | | | |
|---|---|---|---|---|---|---|
| Our firm with supply chain partners offers products that are highly reliable | 1 | 2 | 3 | 4 | 5 | NA |
| Our firm with supply chain partners offers products that are highly durable | 1 | 2 | 3 | 4 | 5 | NA |
| Our firm with supply chain partners offers high quality products to our customers | 1 | 2 | 3 | 4 | 5 | NA |
| Our firm and supply chain partners have helped each other to improve product quality | 1 | 2 | 3 | 4 | 5 | NA |

### INNOVATION
Our firm with supply chain partners ...

| | | | | | | |
|---|---|---|---|---|---|---|
| introduces new products and services to market quickly | 1 | 2 | 3 | 4 | 5 | NA |
| has rapid new product development | 1 | 2 | 3 | 4 | 5 | NA |
| has time-to-market lower than industry average | 1 | 2 | 3 | 4 | 5 | NA |
| innovates frequently | 1 | 2 | 3 | 4 | 5 | NA |

**Section 7. About Your Firm Performance**
As a result of supply chain collaboration, please select the appropriate number that best indicates your firm's overall performance.

| | | | | | | |
|---|---|---|---|---|---|---|
| Market share | 1 | 2 | 3 | 4 | 5 | NA |
| Growth of market share | 1 | 2 | 3 | 4 | 5 | NA |
| Growth of sales | 1 | 2 | 3 | 4 | 5 | NA |
| Return on investment | 1 | 2 | 3 | 4 | 5 | NA |
| Growth in return on investment | 1 | 2 | 3 | 4 | 5 | NA |
| Profit margin on sales | 1 | 2 | 3 | 4 | 5 | NA |
| Overall competitive position | 1 | 2 | 3 | 4 | 5 | NA |

**General Information about Your Firm**
Please give us the following information about your firm for statistical purposes. Please select the appropriate one that best indicates your firm's situation.

1) Number of employees in your company:
    ____ 1 -50    ____ 51-100    ____ 101-250
    ____ 251-500    ____ 501 -1000    ____ Over 1000

2) Average annual sales of your company in millions of $:
    ____ Under 5    ____ 5 to <10    ____ 10 to <25
    ____ 25 to <50    ____ 50 to <100    ____ >100

3) What percentage of your business transactions with your supply chain partners is done electronically?
    ____ Less than 10%    ____ 10-30%    ____ 30-50%
    ____ 50-80%    ____ More than 80%

4) Please indicate the number of tiers across your supply chain.
    ____ <= 3    ____ 4-5    ____ 6-7    ____ 8-10    ____ >10

5) Please mark the position of your company in the supply chain (mark all that applies).
    ____ Raw material supplier ____ Component supplier
    ____ Assembler    ____ Sub-assembler
    ____ Manufacturer    ____ Distributor
    ____ Wholesaler    ____ Retailer

6) Please indicate the SIC category that best describe your primary business:
    ____ Furniture & Fixtures    ____ Rubber & Plastic Products
    ____ Fabricated Metal Products ____ Instruments & Related Products
    ____ Electric & Electronic Equipment ____ Transportation Equipment
    ____ Industrial Machinery & Equipment
    ____ Others (please specify_____)

7) The number of product lines your firm makes_____.

8) Your primary production system (choose most appropriate one).
    ____ Engineer to Order    ____ Make to Order
    ____ Assemble to Order    ____ Make to Stock

9) Your primary manufacturing system (Choose most appropriate one).
    ____ Continuous Flow Process    ____ Assemble Line
    ____ Batch Processing    ____ Job Shop
    ____ Manufacturing Cells ____ Flexible Manufacturing
    ____ Projects (one-of-a kind production)

10) Your present job title:
    ____ CEO/President    ____ Vice President    ____ Director
    ____ Manager    ____ Others (please specify_____)

11) Your present job function (mark all that apply):
    ____ Corporate Executive    ____
    Purchasing/Procurement
    ____ Manufacturing/Operations    ____ Distribution/Warehouse
    ____ Transportation/Logistics    ____ Supply Chain
    ____ Others (please specify _____)

12) The years you have stayed at this organization:
    ____ under 2 years    ____ 2-5 years
    ____ 6-10 years    ____ over 10 years

13) Please rank the importance of the following factors (from 1- most important to 5-least important) in selecting your suppliers (use each number only once).
    ____ Cost    ____ On time delivery    ____ Lead time
    ____ Quality    ____ Delivery reliability

14) If you would like to receive a summary report of the survey results, please provide your mailing address, email, or attach your business card.

**Thank you for your response and time!**

# F: Research Instruments after Large-Scale Study

## IT Resources

*IT Infrastructure Flexibility*

IRIF2    Our systems are compatible
IRIF3    Our systems are scalable
IRIF4    Our systems are transparent
IRIF5    Our systems use commonly agreed IT standards

*IT Expertise*

IRIE1    Our IT staff has good knowledge of information technologies
IRIE2    Our IT staff has the ability to quickly learn and apply new information technologies as they become available
IRIE3    Our IT staff has the skills and experience to develop effective applications and systems
IRIE4    Our IT staff and managers understand our firm's technologies & business processes very well
IRIE6    Our IT staff and managers are knowledgeable about our firm's business strategies, priorities, and opportunities

## IOS Appropriation

*IOS Use for Integration*

IAIG2    Our firm and supply chain partners use IOS for joint forecasting, planning, and execution
IAIG3    Our firm and supply chain partners use IOS for order processing, invoicing and settling accounts
IAIG4    Our firm and supply chain partners use IOS for exchange of shipment and delivery information
IAIG5    Our firm and supply chain partners use IOS for managing warehouse stock and inventories

*IOS Use for Communication*

IAIC2    Our firm and supply chain partners use IOS for conferencing
IAIC3    Our firm and supply chain partners use IOS for message services

IAIC4    Our firm and supply chain partners use IOS for frequent contacts
IAIC5    Our firm and supply chain partners use IOS for multiple channel
         communication

*IOS Use for Intelligence*

IAIL1    Our firm and supply chain partners use IOS for understanding trends in
         sales and customer preferences
IAIL3    Our firm and supply chain partners use IOS for deriving inferences from
         past events (e.g., process exceptions, patterns of demand shifts, what
         worked and what did not work)
IAIL4    Our firm and supply chain partners use IOS for combining information
         from different sources to uncover trends and patterns
IAIL5    Our firm and supply chain partners use IOS for interpreting information
         from different sources in multiple ways depending upon various
         requirements

## Collaborative Culture

*Collectivism*

CCCL1   Our firm and supply chain partners share responsibilities for the
        successes and failures of our working relationships
CCCL2   Our firm considers it as the most normal thing that supply chain partners
        try to cooperate as much as possible
CCCL3   Close cooperation with supply chain partners is to be preferred by our
        firm over working independently
CCCL4   Our firm and supply chain partners focus on joint efforts with a feeling of
        "we are in this together"

*Long Term Orientation*

CCLT1   Our firm wants and expects to have a long-term relationship with supply
        chain partners
CCLT2   Our firm believes that over the long run our relationships with supply
        chain partners are important to us
CCLT3   Our firm believes that short-term inequities in the relationship with
        supply chain partners would be balanced out by mutual benefits over
        the long term
CCLT4   Our firm is willing to make specific investments for long term
        relationships with supply chain partners

*Power Symmetry*

| CCPS1 | Our firm believes that firms in the supply chain have equal influence on each other |
| CCPS2 | Our firm believes that firms in the supply chain that are in a powerful position should meet the needs of less powerful firms in mutually beneficial arrangements |
| CCPS3 | Our firm believes that firms in the supply chain that are in a powerful position should have more to say in their relationships than their partners |
| CCPS4 | Our firm believes that firms in the supply chain that are not in a powerful position should generally follow the will of their partners |

*Uncertainty Avoidance*

| CCUA1 | Uncertain situations in our supply chain are a threat to our firm |
| CCUA2 | Our firm tries to avoid uncertain situations in our supply chain |
| CCUA3 | Our firm tries to avoid unclear and ambiguous situations in our supply chain |
| CCUA4 | Our firm tries to avoid risky situations in our supply chain |

**Trust**

*Credibility*

| TRCR1 | Our supply chain partners are open and honest in dealing with us |
| TRCR2 | Our supply chain partners are reliable |
| TRCR3 | Our supply chain partners respect the confidentiality of the information they receive from us |
| TRCR4 | Our supply chain partners usually keep the promises that they make to us |
| TRCR5 | Our supply chain partners always provide accurate information |

*Benevolence*

| TRBN1 | Our supply chain partners have made sacrifices for us in the past |
| TRBN2 | Our supply chain partners are willing to provide assistance and support to us without exception |
| TRBN3 | Our supply chain partners care for our welfare when making important decisions |
| TRBN5 | We can count on supply chain partners to consider how their actions will affect us |

## Supply Chain Collaboration

*Quality of Information Sharing*

SCIS2    Our firm and supply chain partners exchange timely information
SCIS3    Our firm and supply chain partners exchange accurate information
SCIS4    Our firm and supply chain partners exchange complete information
SCIS5    Our firm and supply chain partners exchange confidential information

*Goal Congruence*

SCGC1    Our firm and supply chain partners have agreement on the goals of the
            supply chain
SCGC2    Our firm and supply chain partners have agreement on the importance of
            collaboration across the supply chain
SCGC3    Our firm and supply chain partners have agreement on the importance of
            improvements that benefit the supply chain as a whole
SCGC4    Our firm and supply chain partners agree that our own goals can be
            achieved through working towards the goals of the supply chain

*Decision Synchronization*

SCDS1    Our firm and supply chain partners jointly plan on promotional events
SCDS2    Our firm and supply chain partners jointly develop demand forecasts
SCDS3    Our firm and supply chain partners jointly manage inventory
SCDS4    Our firm and supply chain partners jointly plan on product assortment

*Incentive Alignment*

SCIA1    Our firm and supply chain partners co-develop systems to evaluate and
            publicize each other's performance (e.g. key performance index,
            scorecard, and the resulting incentive)
SCIA2    Our firm and supply chain partners share costs (e.g. loss on order
            changes)
SCIA4    Our firm and supply chain partners share any risks that can occur in the
            supply chain
SCIA5    The incentive for our firm is commensurate with our investment and risk

*Resource Sharing*

SCRS1    Our firm and supply chain partners use cross-organizational teams frequently for process design and improvement

SCRS3    Our firm and supply chain partners share technical supports

SCRS4    Our firm and supply chain partners share equipments (e.g. computers, networks, machines)

SCRS5    Our firm and supply chain partners pool financial and non-financial resources (e.g. time, money, training)

*Collaborative Communication*

SCCM1    Our firm and supply chain partners have frequent contacts on a regular basis

SCCM2    Our firm and supply chain partners have open and two-way communication

SCCM3    Our firm and supply chain partners have informal communication

SCCM4    Our firm and supply chain partners have many different channels to communicate

SCCM5    Our firm and supply chain partners influence each other's decisions through discussion rather than request

*Joint Knowledge Creation*

SCKC1    Our firm and supply chain partners jointly search and acquire new and relevant knowledge

SCKC2    Our firm and supply chain partners jointly assimilate and apply relevant knowledge

SCKC3    Our firm and supply chain partners jointly identify customer needs

SCKC4    Our firm and supply chain partners jointly discover new or emerging markets

SCKC5    Our firm and supply chain partners jointly learn the intentions and capabilities of our competitors

## Collaborative Advantage

*Process Efficiency*

CAPE1    Our firm with supply chain partners meets agreed upon unit costs in comparison with industry norms

CAPE2   Our firm with supply chain partners meets productivity standards in comparison with industry norms

CAPE3   Our firm with supply chain partners meets on-time delivery requirements in comparison with industry norms

CAPE4   Our firm with supply chain partners meets inventory requirements (finished goods) in comparison with industry norms

*Offering Flexibility*

CAOF1   Our firm with supply chain partners offers a variety of products and services efficiently in comparison with industry norms

CAOF2   Our firm with supply chain partners offers customized products and services with different features quickly in comparison with industry norms

CAOF3   Our firm with supply chain partners meets different customer volume requirements efficiently in comparison with industry norms

CAOF4   Our firm with supply chain partners has good customer responsiveness in comparison with industry norms

*Business Synergy*

CABS1   Our firm and supply chain partners have integrated IT infrastructure and IT resources

CABS2   Our firm and supply chain partners have integrated knowledge bases and know-how

CABS3   Our firm and supply chain partners have integrated marketing efforts

CABS4   Our firm and supply chain partners have integrated production systems

*Quality*

CAQL1   Our firm with supply chain partners offers products that are highly reliable

CAQL2   Our firm with supply chain partners offers products that are highly durable

CAQL3   Our firm with supply chain partners offers high quality products to our customers

CAQL4   Our firm and supply chain partners have helped each other to improve product quality

*Innovation*

CAIN1    Our firm with supply chain partners introduces new products and services to market quickly

CAIN2    Our firm with supply chain partners has rapid new product development

CAIN3    Our firm with supply chain partners has time-to-market lower than industry average

CAIN4    Our firm with supply chain partners innovates frequently


*Firm Performance*

FP3      Growth of sales

FP4      Return on investment

FP5      Growth in return on investment

FP6      Profit margin on sales

# Index